HOME BUILDER'S GUIDE TO SEISMIC RESISTANT CONSTRUCTION

FEDERAL EMERGENCY MANAGEMENT AGENCY

Fredonia Books
Amsterdam, The Netherlands

Home Builder's Guide to Seismic Resistant Construction

by
Federal Emergency Management Agency

ISBN: 1-4101-0879-1

Copyright © 2005 by Fredonia Books

Reprinted from the 1998 edition

Fredonia Books
Amsterdam, The Netherlands
http://www.fredoniabooks.com

All rights reserved, including the right to reproduce this book, or portions thereof, in any form.

TABLE OF CONTENTS

PREFACE

SECTION		PAGE
1.	INTRODUCTION, BACKGROUND AND SEISMIC RISK AREAS	1
2.	PRINCIPLES OF SEISMIC RESISTANCE IN DWELLINGS	3
3.	ARCHITECTURAL CONSIDERATIONS	9
4.	SITE SELECTION	12
5.	ELEMENTS OF THE SEISMIC RESISTANT SYSTEM	17
6.	FOUNDATIONS AND FOUNDATION DETAILS	21
7.	FLOORS	27
8.	SHEAR WALLS	32
9.	ROOF REQUIREMENTS	43
10.	MASONRY CHIMNEYS	49
11.	CONCRETE MASONRY	52
12.	CLAY MASONRY	58
13.	MASONRY AND STONE VENEER	59
14.	BUILDING CODE AND REFERENCES	60
15	HOME BUILDERS CHECK LIST	73
16.	APPENDIX: TYPICAL REGULAR FLOOR PLANS FOR EARTHQUAKE RESISTANCE	A1

ACKNOWLEDGMENTS

PREFACE

One of the primary goals of the Federal Emergency Management Agency (FEMA) and the National Earthquake Hazards Reduction Program (NEHRP) is to encourage design and building practices that address the earthquake hazard and minimize the resulting damage to both life and property. The cornerstone of FEMA's program to accomplish this goal is *mitigation* — that is, to strengthen the building and all of its' components against the force of the earthquake *before* it strikes.

This Guide updates, revises, and replaces the FEMA-232 Document, dated July 1992, which was based on "The Home Builder's Guide for Earthquake Design", published by the Department of Housing and Urban Development (HUD) in June 1980.

FEMA wishes to express its gratitude to the contractor, "SOHA Engineers", and the many subcontractors, contributers, and reviewers that were involved in this effort. It was their hard work and dedication that resulted in the successful completion of this guide.

Federal Emergency Management Agency

1. INTRODUCTION, BACKGROUND AND SEISMIC RISK AREAS

A. INTRODUCTION

The primary purpose of this Home Builders Guide to Seismic Resistant Construction, hereafter referred to as the "Guide", is to encourage homeowners and builders of one and two family residences to employ construction practices intended to provide resistance to damage from earthquakes. This Guide can be used as a convenient resource for gaining an understanding of the basic principles of seismic resistant construction. The Guide presents a discussion of how earthquake forces impact conventional residential construction. A discussion is included on how basic structural components can be assembled to achieve earthquake resistance and how essential features such as foundations, walls, floors and roofs interact to resist earthquakes. Warnings are included regarding special requirements for easily damaged components and configurations such as masonry chimneys, masonry veneers, split level plans and floor plan irregularities.

By using information available in the Guide, residential units can be built with structural features that are positioned, dimensioned, constructed and interconnected properly to resist earthquakes. The Guide incorporates lessons learned from the 1994 Northridge and other recent earthquakes. Application of the principles contained in the Guide should improve the overall quality of the home building process and will result in better performance of residential buildings.

Limitations on the application of information presented in the Guide must be carefully reviewed. The Guide is NOT a substitute for seismic resistant design provisions of the applicable local Building Code. The Guide may not be used in place of the Building Code. It is a reference document providing details of construction some of which may not be specifically described or presented in the Code.

In regions where earthquake resistance is of interest, but seismic resistant construction practices are not required by code or regulation, the Guide can provide direction to builders. Also, in some instances, the Guide will recommend construction practices that are more conservative than code requirements. These are usually the result of lessons learned from the Northridge and other earthquakes. Following the recommended practices in this Guide should result in better performing houses. The Guide may not apply to conditions found in the design and construction of houses that do not conform to conventional configurations and construction practices. Specific reference to unusual conditions are noted in the Guide. Professional guidance in designing earthquake resistant houses should be sought where houses do not conform to conventional configurations.

Details should not be selected at random from the Guide. A complete earthquake resistance system requires all the necessary elements to be present. Random selection of details without consideration of the building as a complete system may result in ineffective earthquake resistance.

To use the Guide properly, a builder should first attempt to understand the principles of seismic resistance as they are presented in the Guide. Once the principles are understood, the importance of following the information provided in the Guide will be apparent.

B. BACKGROUND

Three model building codes are available for use by home builders in specific areas of the country. In the western states, the Uniform Building Code (UBC) published by the International Conference of Building Officials (ICBO) is used by most government agencies regulating building construction. Most southern states use the Standard Building Code (SBC) sponsored by the Southern Building Code Congress International (SBCCI). The northeast and parts of the midwest use the National Building Code (NBC) issued by the Building Officials and Code Administrators, International (BOCA). Local codes may apply in certain areas. The three model codes noted above will be replaced by the International Building Code (IBC) scheduled to be released in the year 2000.

The UBC includes prescriptive requirements that apply to residential construction. The other two model Codes refer to the CABO One and Two Family Dwelling Code published by the Council of American Building Officials. This Code is a compilation of data from the model codes and standardizes the requirements for home construction in a single document. The CABO Code is scheduled to be replaced by the International Residential Code (IRC) in the year 2000. The purpose of the model codes, as stated in the UBC, is to provide minimum standards to safeguard life or limb, health, property and the public welfare by regulating the design construction, quality of materials, use, occupancy, location and maintenance of all buildings and structures.

To encourage a national awareness of the hazards of earthquakes, the Federal Emergency Management Agency (FEMA) initiated the preparation of this document.

The Guide is not intended to supplant the CABO One and Two Family Dwelling Code or the three model codes, but rather to offer explanations, advice and construction details for use when constructing a home to provide for resistance to earthquakes. It is a reference to be used by home builders and others to help in understanding how earthquakes cause damage and how homes can be constructed with increased resistance to earthquakes.

In certain areas of the country, other natural hazards such as wind and flood may dictate structural requirements. An analysis of these hazards and their mitigation requirements should be made to compare the recommendations of this Guide with the requirements for other hazards to determine whether construction to mitigate one hazard will satisfy the others.

C. SEISMIC RISK AREAS

Details presented in the Guide are suggested for three areas of earthquake activity and severity in the United States. These three areas are referred to in the Guide as high, moderate and low seismic risk areas. This designation of seismic activity may differ from the methods used by the three model codes and CABO to define areas or zones of seismic activity and risk. See Table No. 5 on page 65 for a comparison of CABO One and Two Family Dwelling Code, SBCCI, BOCA and UBC risk designations and the seismic risk areas used in the Guide. For purposes of this Guide, high, moderate and low seismic risk areas can be determined from a map taken from the Uniform Building Code, see page 65, and comparing the CABO and UBC zone designations with the Seismic Risk Areas indicated in Table 5, page 65.

When referring to this map, the home builder should be aware that the seismic risk areas shown may differ from similar geographical areas or seismic zones which apply locally. If doubt exists as to whether the location is in a high, moderate or low seismic risk area, the local building official should be consulted.

2. PRINCIPLES OF SEISMIC RESISTANCE IN DWELLINGS

Earthquakes result in ground motions, both horizontal and vertical, which can be compared to waves. The motion is generally vibratory and will cause a structure to move rapidly first in one direction and then another.

Earthquakes generate internal forces in a structure due to inertia. Inertia can be described as the tendency of a body at rest to remain at rest and a body in motion to remain in motion. See Figure 1. The internal forces depend on the direction of ground motion caused by an earthquake and act side to side (horizontal) and/or up and down (vertical). The more pronounced earthquake forces are usually horizontal, i.e., lateral forces acting back and forth parallel to the ground. Because the ground motion moves back and forth, the effects of inertia cause a building to be distorted and can result in severe damage. The effects of vertical acceleration are normally considered to be offset by the building weight and will only cause damage in unusual situations.

The sketches shown in Figure 2 illustrate the effects of inertia on a simple structure caused by back and forth motion parallel to the ground. Similar effects would occur if the ground were stationary and a horizontal force was applied in a back and forth manner at the roof line. Earthquake effects are usually represented in this manner. The objective of earthquake resistant construction is to resist the effects of ground movement.

In the remaining discussion, applied forces parallel to the ground (horizontal or lateral) will be considered as representative of earthquake forces.

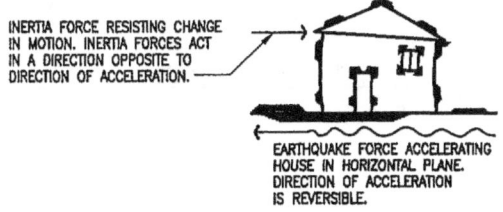

Figure 1

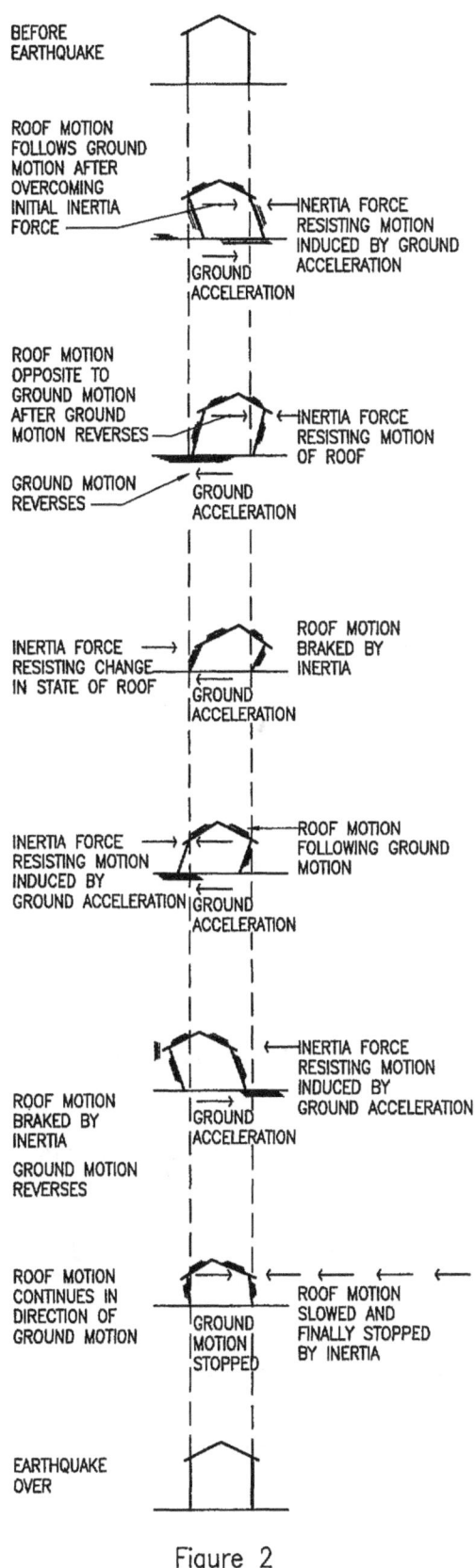

Figure 2

In discussing earthquake resistance in subsequent sections, engineering terminology will be used that may not be familiar to the user of the Guide. The most frequently used terms are:

Shear — the tendency of a force applied to one part of an element to cause sliding relative to the remainder of the element along a plane parallel to the force. See Figure 3.

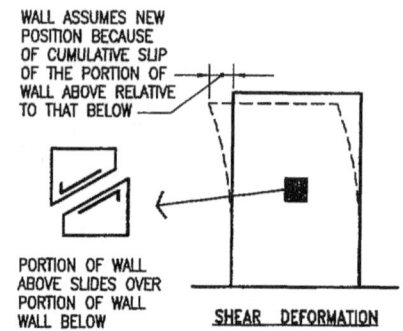

Figure 3

Tension — the force causing the pulling apart of one element from another or causing elongation of a material. See Figure 4.

Compression — the force causing pushing of one element on another or causing shortening of material. The opposite of tension. See Figure 5.

Chord — used as a synonym for flange of a diaphragm or shear wall.

Collector — a member (usually wall top plates) used to accumulate forces from a horizontal diaphragm along the portion of its length where no shear wall exists and delivers the forces to a shear wall or another diaphragm. See Figure 11.

Diaphragm Ratio — The ratio of the length to width of a horizontal diaphragm. See Figure 6.

Aspect Ratio — The ratio of the height to width of a shear wall. See Figure 6.

Deflection/Drift — The displacement of the top of the wall relative to the fixed bottom of the same wall.

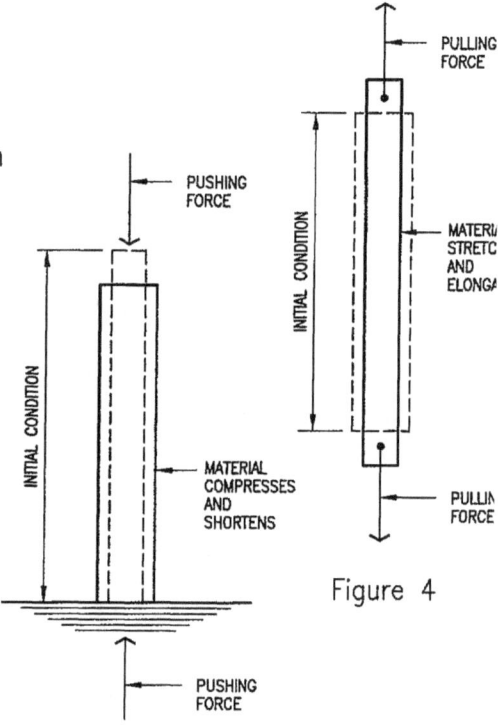

Figure 4

Figure 5

Conventionally constructed dwellings can resist earthquake forces because of their box-like configuration. To be effective, the box-like configuration must be complete and well tied together, with mostly square or rectangular floors and roofs and solid wall panels on all four sides of the building. This is presented in Figure 7.

Roofs and framed floors are known as horizontal diaphragms. Floors are usually flat and in the same plane. Roofs may be flat, pitched or gabled. Each roof configuration can be made to function effectively as a horizontal diaphragm as shown in Figures 8 and 9.

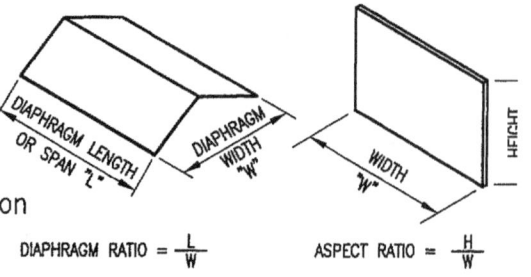

Figure 6

(4)

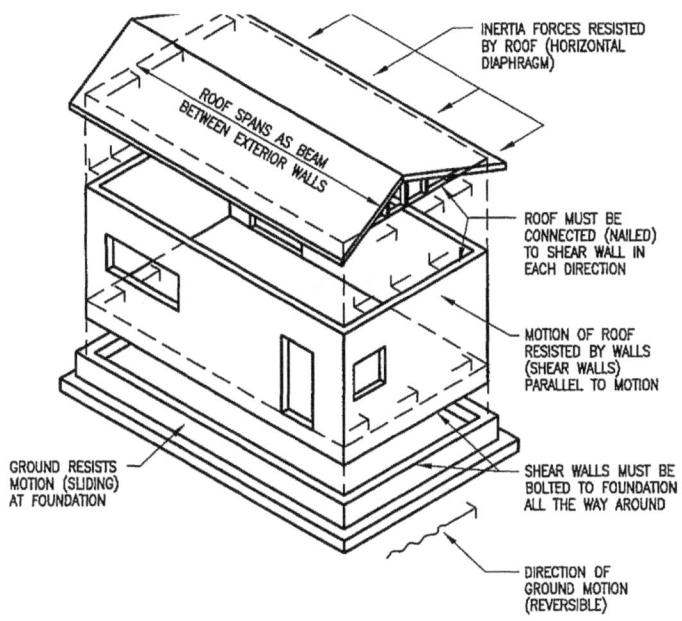

Figure 7

A horizontal diaphragm can be compared to a steel wide flange beam laid on its side with the web oriented in a horizontal plane and the flanges in a vertical plane. In residential construction, the exterior wall top plates act as flanges and the roof sheathing functions as the web. The flanges (top plates) carry tension or compression and the roof sheathing transmits the shear stresses caused by the earthquake.   See Figures 9 and 10.

Those walls resisting the horizontal forces are known as shear walls or, in the case of a resisting element being a portion of a longer wall, as a shear panel. See Figure 7.

Shear walls or panels can be pictured as upright beams with the end studs of the sheathed portion acting as the flanges and the sheathing between end studs as the web. Thus, shear walls are the same as vertical cantilever beams (a beam supported at one end but not at the other) fixed at the foundation with the top of the wall being loaded by the roof diaphragm. See Figure 15.

Interior and exterior walls can resist the horizontal forces transmitted by the roof and floor diaphragms. Since earthquakes can create forces in any direction, opposite and parallel pairs of walls resist loads in a single direction while each of the exterior walls participates in resisting rotation or torsional distortion of the house. See Figure 11.

If shear walls, properly constructed and anchored, are located on each of four sides of a square or rectangular building and there is a floor and roof that can function as a diaphragm in either direction, then the building functions as a box. See Figure 7.

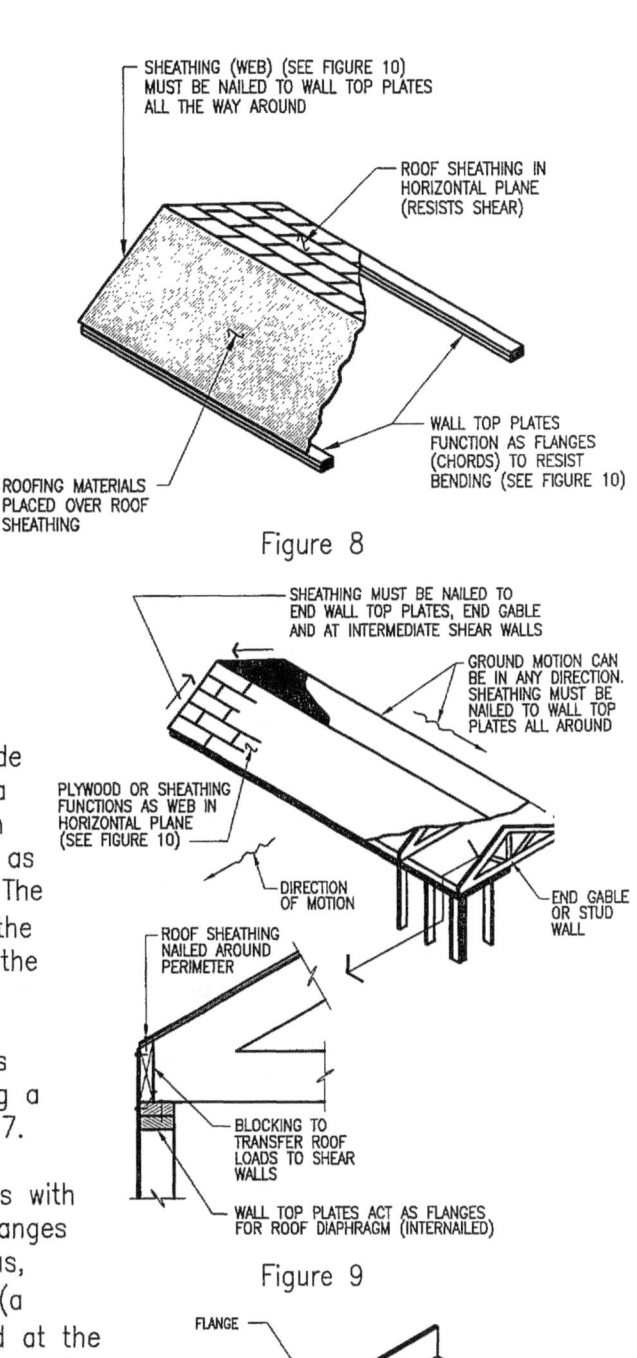

Figure 8

Figure 9

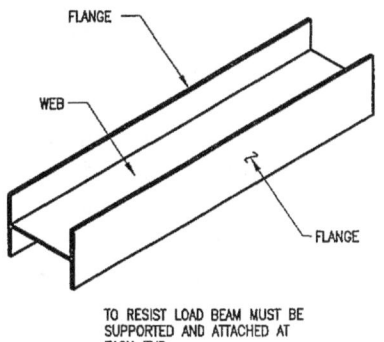

TO RESIST LOAD BEAM MUST BE SUPPORTED AND ATTACHED AT EACH END.

Figure 10

Provided that the horizontal diaphragms and shear walls are appropriately sized and constructed, forces introduced into a building by an earthquake can be resisted by the box configuration and damage to the building will be minimized.

In order for the box to function properly, the horizontal diaphragms and shear walls must be constructed to resist the induced forces without collapsing the box. It is essential that the horizontal diaphragm be connected to the shear walls and the shear walls be securely fastened to the foundations. See Figure 12.

In structures such as framed residential buildings, where connections of walls to foundations are adequate to prevent sliding, deflections or drift is the primary cause of damage. In high seismic risk areas, greater consideration should be given to increasing the overall stiffness of a house to control damage to brittle materials such as stucco, gypsum board finshes, interior furnishings, masonry chimneys and veneer, all of which are especially vulnerable to damage from drift. Limiting the height to width ratio of shear walls to two to one (2:1) in high and moderate seismic risk areas, and/or providing more shear wall width than is needed to meet minimum requirements of the building code will contribute to stiffness. Providing the correct location and adequate number of bolts for attaching sills to the foundations will prevent sliding and help to control damage.

Another effect of earthquake forces is to cause overturning of walls or shear panels. This tendency to rock on the foundation can be resisted by anchors that will hold the wall to the foundation, commonly referred to as hold-downs. With reference to Figure 13, as the ground accelerates to the left the inertia forces act in the opposite manner, to the right. Hold-down anchors should be inserted at the ends of shear walls to resist rocking. Since earthquake forces oscillate back and forth, hold-downs should be placed at each end of a shear wall or panel. See Figures 13 and 14.

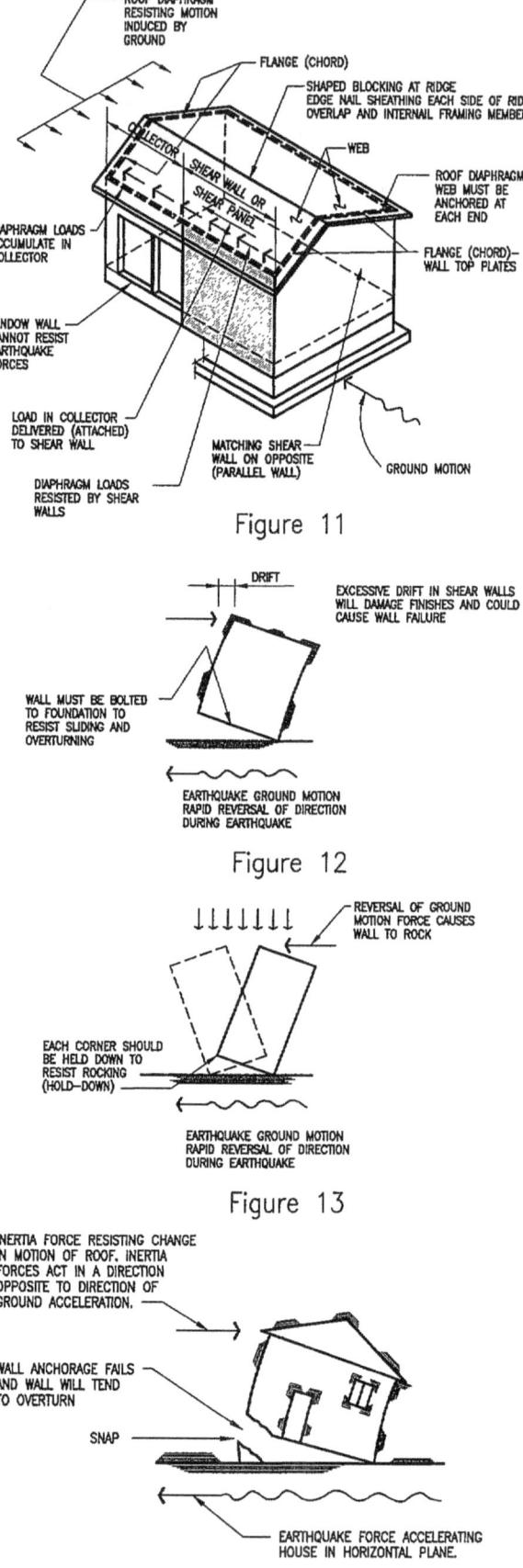

Figure 11

Figure 12

Figure 13

Figure 14

Earthquake forces in shear walls are transferred to the walls and back to the ground through the foundation. For one and two family dwellings, a continuous foundation located under the perimeter walls and interior shear walls is most commonly used. See Figure 16.

Sheathing of diaphragms and shear walls resists earthquake induced shear forces. Several types of sheathing are available. Plywood*, oriented strand board*, gypsum board and stucco are often used on wood framed shear walls. Plywood and oriented strand board are most effective. See Figure 15. All sheathing must be fastened to the framing. Nails are most often used. See Figure 17. The type of nails used can vary, but common or box nails are preferred for plywood and oriented strand board. Cooler nails are most frequently used with gypsum board. Staples have not proved to be universally satisfactory for attaching plywood or oriented strand board and should be avoided in high and moderate seismic risk areas unless proven installation methods are followed for the specific use.

The degree of earthquake resistance required depends on the location and risk factor associated with that particular region of the country. The Guide uses high, moderate and low seismic risk area designations. The map on page 64 used together with Table 5 on page 65 indicates locations of the seismic risk areas.

* Wood Structural Panel

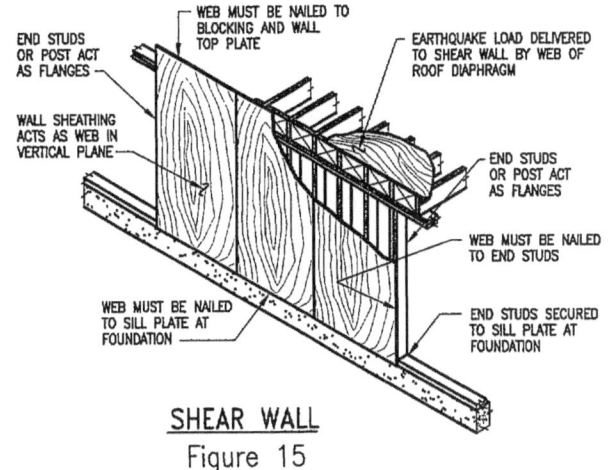

SHEAR WALL
Figure 15

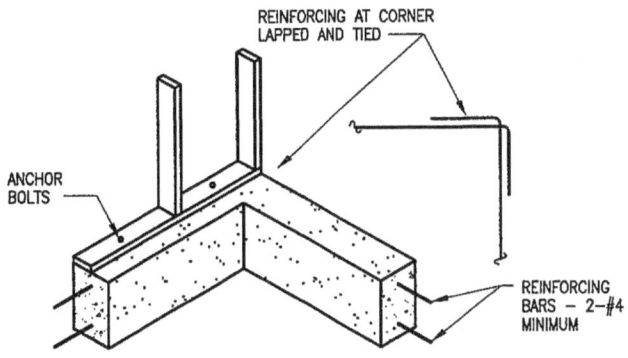

CONTINUOUS FOUNDATION
Figure 16

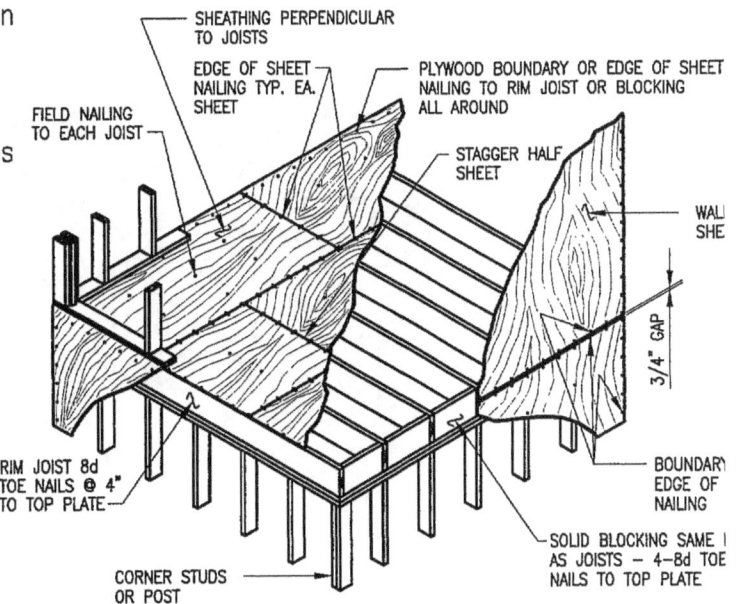

TYPICAL DIAPHRAGM
Figure 17

In the past, home builders have used all wood products for the construction of one and two family houses. Most builders are thoroughly familiar with the materials and construction practices for conventional wood construction. With growing concerns for the depletion of the country's natural resources and environmentalists' concern for clear cutting forests, cold-formed steel framing is coming into greater usage. Homes framed in steel employ steel studs and roof trusses made of cold-formed steel members or joists or rafters also fabricated from cold-formed steel material. Connections of framing members are usually made with self-tapping sheet metal screws. Conventional wall covering materials such as gypsum board and stucco are applied to the studs in a manner similar to that for wood framing. Because cold-formed steel framing for residential construction is not yet widely used, building codes have not adopted conventional construction requirements for this material. In some areas of the country using the UBC, cold-formed steel framing must be designed by a design professional.

Cold-formed steel construction for seismic resistance is similar to that for wood framing. Steel stud framed shear walls can be braced with light-gage steel diagonal straps fastened with self tapping sheet metal screws. The straps are usually positioned in an "X" pattern. See Figure 18 The building official having jurisdiction should be consulted as to whether the use of plywood sheathing over steel studs is allowed for seismic resistance.

As in wood framed construction, hold-downs are usually required at each end of a shear wall. Hold-downs, in general, are steel strap type anchors with one end buried in the foundation and the projecting strap screwed to the edge of the steel studs in the end of the wall.

Figure 8

3. ARCHITECTURAL CONSIDERATIONS

The Guide has been developed for application to simple, conventional plans with regular and typical architectural features, see Appendix A. The Guide is most useful for dwellings with the following characteristics.

1) Regular or symmetrical floor plans.

2) Floor plans with balanced widths of shear wall or bracing in each exterior wall.

3) Elevations with limited openings.

4) Roof shapes that are symmetrical with minimal offsets and limited openings.

5) Conventional wall framing systems. Typically these are wood stud walls, steel stud walls, clay masonry walls and concrete masonry walls.

6) Chimneys located and anchored as described in the Guide.

Example floor plans indicating some of the most beneficial features for making residential buildings resistant to earthquake shaking are shown in the Appendix. These plans represent regular and reasonably symmetrical configurations. Variations of these plans that incorporate most of the features of the plans shown in the Appendix are appropriate for use with the Guide.

Architectural expression in dwellings may not be limited to conventional floor plans and exterior elevations. The guidelines and recommendations for earthquake resistant construction formulated for conventional types of dwellings may not apply to irregular or unusual structures. In general, however, adoption of the details and procedures outlined in this Guide should increase the ability of any building to resist earthquake forces. Some irregular and unusual structures may have details and configurations that are especially vulnerable to earthquake damage particularly in high and moderate seismic risk areas. Methods to specifically deal with these details and configurations are not covered in the Guide.

The following dwelling types and configurations listed below <u>are not</u> included in this Guide. Table No. 1 on page 11 illustrates 5 common examples.

1) Platform framing on stilt supports.

2) Post and beam framing with long, uninterrupted expanses of glass.

3) Overhanging two story construction where the exterior walls do not continue to the ground.

4) Staggered floor systems involving more than two levels.

5) Exterior glass walls without shear panels

6) Floors and roofs with extreme overhangs and balconies.

7) Pole-supported framing.

8) Attached carports with open sides and corner posts.

9) Open courtyards with all glass building wall enclosures.

10) Buildings constructed on slopes steeper than 3:1.

It is possible to build earthquake resistive dwellings which have configurations and features such as those described above. However, the Guide is not appropriate for use with such dwellings without consultation with design professionals familiar with earthquake resistant design. Many of the details included in the Guide may be helpful to design professionals.

Split-level construction is defined as construction having two floor levels separated by a vertical offset that interrupts the planes of the floor or roof diaphragms. Figure 19 illustrates an offset at the floor and roof levels.

Split-level houses require special details at the plane of intersection of the floors and roof with the wall. Details are included in the Guide to provide for connecting elements together to resist earthquakes. Split-level homes can be particularly vulnerable to damage in high and moderate seismic risk areas unless appropriate details as presented in the Guide are used. Figure 20 shows an example of the type of detail needed. See Section 7 for additional discussion of split-level construction.

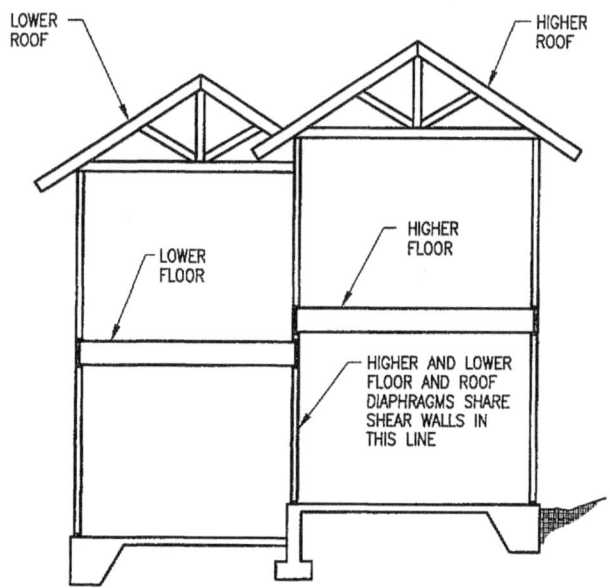

Note
See Split-Level
Floor Tie Details
on Page 31

Figure 19

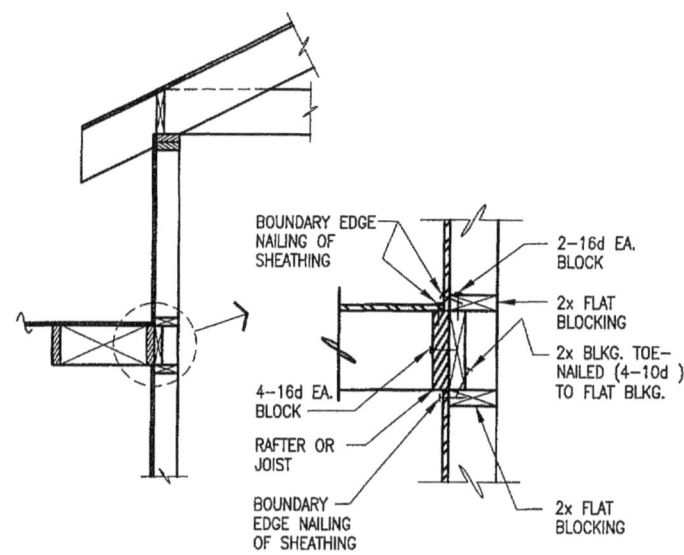

SPLIT-LEVEL TIE FRAMING PERPENDICULAR TO SHEAR WALL

Figure 20

Table No. 1 Architectural Design Exclusions	
Platform Framing on Stilt Supports	STILT STRUCTURE — DIAGONAL BRACING ONLY
Pole Supported Houses	ISOLATED WOOD POLES
Long, Uninterrupted Exterior Glass Walls	OPEN FRONT BUILDINGS WITH LARGE GLASS EXPOSURE
Post and Beam Framing with All Glass Exterior Exposure	
Overhanging Two Story Construction (Corner Columns Do Not Continue to Ground)	PEDESTAL TYPE STRUCTURE, OFF FRAMING OR LAF CANTILEVERS

4. SITE SELECTION

Because the motion of the ground induces earthquake lateral forces in buildings, it is desirable, but not always possible, to build on sites located on stable and solid geologic formations. Deep and unbroken rock formations, referred to as bedrock, generally will minimize earthquake damage. It is not always possible to restrict the location of buildings to such geologically stable sites. However, by adhering to good construction practices such as those illustrated in this Guide, earthquake resistive houses can be built on other more commonly encountered site conditions.

Houses built on unstable, questionable terrain may by vulnerable to damage due to ground failure as well as earthquake shaking in any seismic risk area. Greater care is required in high and moderate seismic risk areas subject to frequent and/or strong earthquakes, especially where unusual ground conditions have previously caused damage to buildings during an earthquake. The principal hazards are landsliding, liquefaction, settlement or subsidence, and surface fault rupture.

Landsliding

Landsliding is a potential hazard on hillside sites. A landslide scarp is shown in Figure 21 and damage to houses due to landsliding is shown in Figure 25 on page 15. Landslide-prone hillside sites include those that are underlain by previous landslides and those that, while presently stable, are located in unstable geologic formations. Landsliding is a particular hazard where poorly compacted fill has been placed on a hillside slope. Landsliding may cause damage in the absence of earthquake shaking and should be of concern even in low seismic risk areas.

Rockfalls or falling boulders are hazards that are similar to landsliding. If an unstable rock slope is present above a dwelling, boulders could be dislodged by earthquake shaking and impact the dwelling below.

Stable hillside sites with slopes of three to one (3:1) or less are generally satisfactory providing there is no previous history of landslide or movement on similar slopes in the vicinity and the slope does not contain poorly compacted fill. Any hillside site for which a professional soil consultant has made a favorable evaluation of the slope or designed the slopes to be stable can be usually regarded as satisfactory.

Figure 21

Compacted fills placed on slopes with engineering design and control during construction may also be regarded as stable. See Figure 23. For steep sites, geotechnical and engineering assistance is recommended in all seismic risk areas.

Liquefaction

Liquefaction is a phenomenon where loose saturated sandy soil loses bearing capacity and becomes like quicksand during earthquake shaking. The soil can spread laterally or may lose the ability to provide support for building foundations. Young alluvial deposits near rivers, lakes, or bays, or reclaimed land formed by placing uncompacted fills in lakes or bays are most likely to exhibit liquefaction potential.

Settlement

Any site on which fill has been placed without professional control should be regarded as suspect and possibly subject to either long term or earthquake induced settlement or subsidence. Settlement may be most pronounced in high seismic risk areas. The potential for settlement should be carefully evaluated, however, even in low seismic risk areas.

Surface Fault Rupture

Dwellings should not be located directly on an active earthquake fault due to the possibility of surface fault rupture and abrupt shearing displacement occurring along the fault during an earthquake. To avoid construction directly on a fault, a zone of at least 100 feet (50 feet on either side of the fault) where no construction is allowed should be observed. In the event of a large or moderate earthquake a dwelling situated directly on an active fault or in a fault zone would be subject to severe damage even if it was well constructed. Figures 22 and 25 illustrate the types of displacement that can occur along earthquake faults. Damage to two houses that were situated on faults during two different earthquakes are shown in Figure 26.

Site Evaluation

Before selecting a building site, the builder should investigate the existence of any known or recognized geotechnical or geologic hazards discussed above. Maps are available for some locales from the US Geological Survey (USGS), state geologic agencies or from county and city agencies that show locations of active earthquake faults, fault zones and areas of potential landsliding and liquefaction. Some local jurisdictions have used these maps to delineate special zones where residential construction could be vulnerable to damage in the event of an earthquake.

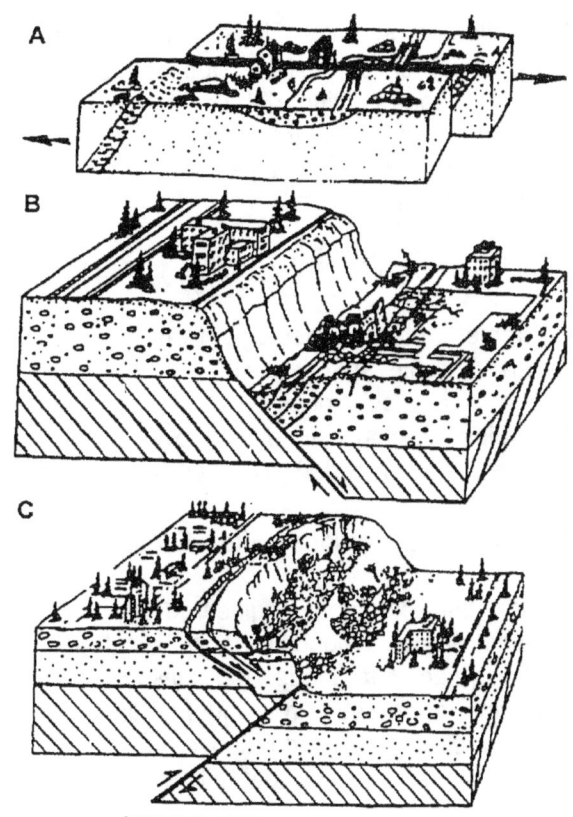

Schematic Diagrams of Surface Fault Displacement. A) Strike Slip. B) Normal Slip. C) Reverse Slip

Figure 22

In areas suspected of being exposed to geotechnical hazards, maps or information indicating such conditions should be sought. A professional consultant may be retained to make a site-specific assessment if a hazardous condition is known or suspected to exist at a site. Construction in areas of low seismic risk may not justify an extensive investigation.

When site conditions are unstable or of questionable soundness, such sites should be evaluated by a professional consultant. If a hazardous condition exists, the site may still be suitable if special measures are taken for the site and/or for the construction of the dwelling. If the site conditions can be sufficiently improved, an earthquake resistive house can be built using the information and illustrations shown in the guide.

Summary

The following summary lists Qualified and Disqualified (hazardous) building sites. Qualified sites are stable sites for which the building details in this Guide are applicable. Disqualified sites require further evaluation as discussed above. Some of these types of sites are shown in Figures 23 and 24.

Qualified Sites For Which Guide is Applicable

1. Stable and solid geologic formations – bedrock
2. Firm, stable soil deposits
3. Stable hillside slopes
4. Engineered land fill placed over stable soils
5. Sites recommended by a professional soil consultant.

Disqualified (Hazardous) Sites Requiring Further Evaluation

1. Sites built on non-engineered fill
2. Sites prone to landsliding or rockfalls
3. Reclaimed waterlands, marshes, or alluvial soil sites having a potential for liquefaction.
4. Sites directly over an active fault or within a 100 feet wide zone, i.e., 50 feet either side of the fault.

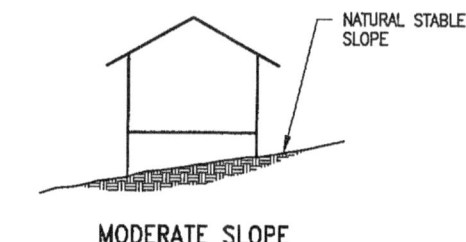

MODERATE SLOPE

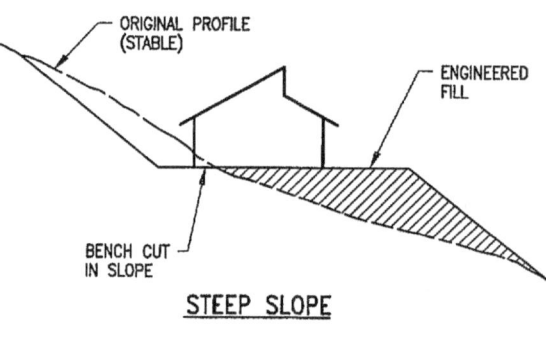

STEEP SLOPE

QUALIFIED

Figure 23

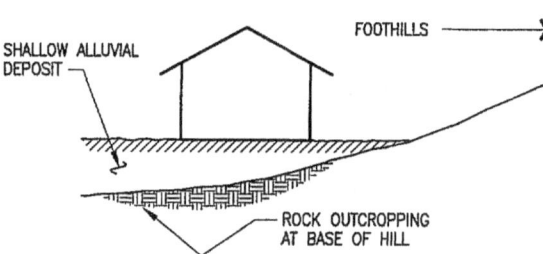

UNCONSOLIDATED FILL

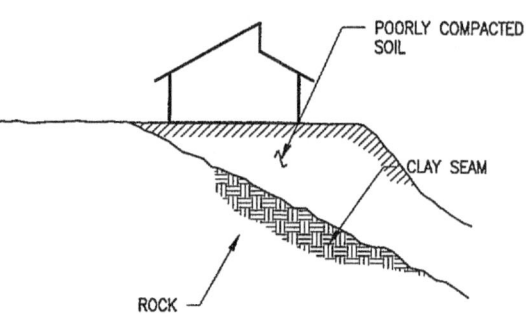

UNSTABLE SLOPE

DISQUALIFIED

Figure 24

Figure 25

Figure 26

5. ELEMENTS OF THE SEISMIC RESISTANCE SYSTEM

This section identifies and describes the components of residential construction that resist earthquake forces. The text and illustrations show how a complete, uninterrupted and well defined load path is constructed. It is essential that all of the seismic resisting components be tied together to form an earthquake resisting system without weak links. Connections between the resisting elements provide the continuity necessary to assure an uninterrupted path. The three principal elements of the system are the floor and roof diaphragms, shear walls and the foundation. See Figure 27.

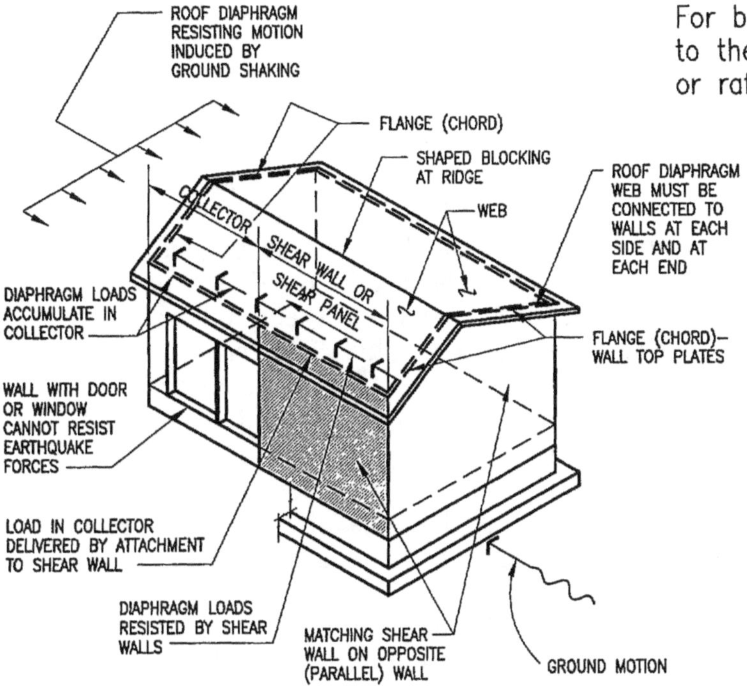

The function of the floor and roof diaphragms is to transfer earthquake forces to shear walls which are usually located at the exterior of the building. For wood framed houses, diaphragm framing usually consists of the floor or roof sheathing, joists or rafters, blocking at the ends of joists or rafters and wall top plates. Cold formed steel framed houses are framed in essentially the same manner.

Wood Structural Panel is always acceptable for roof and floor sheathing. Diagonal sheathing, although not commonly used, but still permitted by Code, can be used instead although not as effectively. Straight sheathing may be used in areas of low seismic risk but should be avoided in high and moderate seismic risk areas. Spaced roof sheathing, often used for shingled roofs, is not recommended but may be used in low seismic risk areas where wind loads are also low.

For best performance, sheathing should be applied to the framing with face grain across the joists or rafters and with nailing on all edges and at intermediate supports. See Figure 17 on page 7. Each sheet must be nailed along each short edge to a framing member. Abutting short edges of adjoining sheets must be nailed to the same (mutual) framing member. The long edge of the sheet may be nailed to blocking inserted between the joists or rafters with abutting edges nailed to the same member. Blocking of edges to resist lateral forces is not usually required for roof or floor diaphragms in residential construction, but can be used to increase the stiffness and strength of the diaphragm. Where tongue and groove sheets are used, blocking may be omitted along the long edges. A bead of construction adhesive should be used on the top edge of floor joists. This is usually done to prevent squeaking but tests have shown that it also increases stiffness and improves performance. Adhesive on roof framing members can be used to stiffen the roof diaphragm.

Figure 27

Edge of sheet nailing of floor and roof sheathing is required at diaphragm boundaries, along exterior wall plate lines and along interior walls used as shear walls. Sheathing is usually edge nailed to blocking or rim joists or similar members attached to the wall top plates. See Figure 17 on page 7.

Nailing of floor and roof sheathing prescribed in building codes for conventional construction is ordinarily adequate for diaphragm strength.

Exterior wall top plates function as flanges or chords for the diaphragm and resist the tension and compression resulting from beam action. The wall double top plates at the boundaries of the diaphragms must be spliced along the widths and lengths (perimeter) of the diaphragms in order for them to be "continuous". See Figure 28.

Because of strength and stiffness limitations, diaphragms are restricted to certain diaphragm ratios of length (L) to width (W). See Table No. 3 on page 39. If the horizontal distance (span) between exterior walls is too large, the diaphragm becomes long and narrow and consequently too flexible. The diaphragm ratio (L/W) is important because excessive deflection of the diaphragm may cause wall damage and glass breakage. If the distance between exterior walls is too great, the proper diaphragm ratio may be achieved by using interior shear walls parallel to the exterior walls to reduce the diaphragm span.

The primary vertical resisting elements are the shear walls that support the floor and roof diaphragms and resist earthquake forces generated in the diaphragms. See Figure 11 on page 6. Aspect ratios of shear walls (vertical diaphragms) are established to control drift. The recommended height (H) to width (W) ratios shown in Table 3 on page 39, should not be exceeded and spacing of shear walls should be as prescribed in the building code or as shown in this Guide.

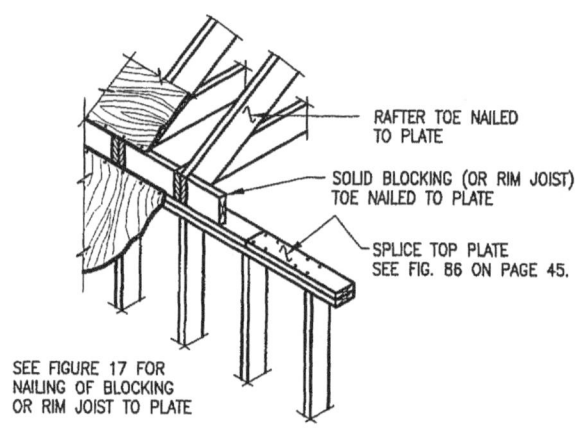

TOP PLATE SPLICE

Figure 28

Shear wall framing usually consists of wall top plates, studs, sheathing, floor sill plate, hold-downs and other connections. Top plates should be doubled, lapped and spliced for continuity. Studs should be doubled at the ends of shear walls and internailed stud to stud. Sheathing must be nailed to blocking or rim joists attached to the wall top plates, to the end or corner studs, and to the floor plate or sill. The floor sill plates must be bolted to the foundation or otherwise attached by special hardware designed to substitute for bolts.

Several sheathing materials are available with Wood Structural Panel sheathing being the most effective. See Table No. 3 on page 39. It is important to use Wood Structural Panel sheathing in high seismic risk areas.

the studs can also be used to resist earthquakes. See Table No. 3 on page 39. As demonstrated in the Northridge and other earthquakes, when gypsum board is used for shear wall sheathing in high seismic risk areas severe damage can result from strong earthquake shaking. Gypsum board is usually adequate as sheathing on shear walls in moderate and low seismic risk areas. Gypsum board may be applied to the walls with long edges vertical or horizontal. But if long edges are horizontal they should be blocked.

Stucco over wire lath applied directly to the studs can be used for seismic resistance in all seismic risk areas, although, it is not advised in high seismic risk areas. The lath must be securely nailed to the studs, wall top plates and sill. In high seismic risk areas, stucco can be vulnerable to severe cracking as demonstrated in the Northridge and other earthquakes.

Let-in bracing provides only minimal seismic resistance and should not be relied on except in low seismic risk areas where wind loads are also low.

For shear walls to be effective in resisting the diaphragm loads and controlling damage they should have height to width ratios in high and moderate seismic risk areas of 2:1 or less. See Figure 29. A larger ratio may be allowed by certain codes, but experience in earthquakes has demonstrated that limiting height to width ratios will improve performance. Shear walls, in two story construction need only satisfy this ratio in each story although better performance can be achieved by computing the ratio using the full height of the wall. A minimum width of 4'-0" for shear walls in high and moderate seismic risk areas is advised.

As noted before, earthquake forces must be transferred from the roof diaphragm into the shear walls and through the walls to the ground at the foundations. At the interface of the horizontal diaphragms and shear walls, the roof and/or floor sheathing rim joists, and the wall sheathing is usually nailed to the vertical face of the same block or rim joists. Where the wall sheathing only extends to the top of the top plates, the blocking or rim joist must be securely nailed or connected to the top plates. See Figure 30 and Figure 32.

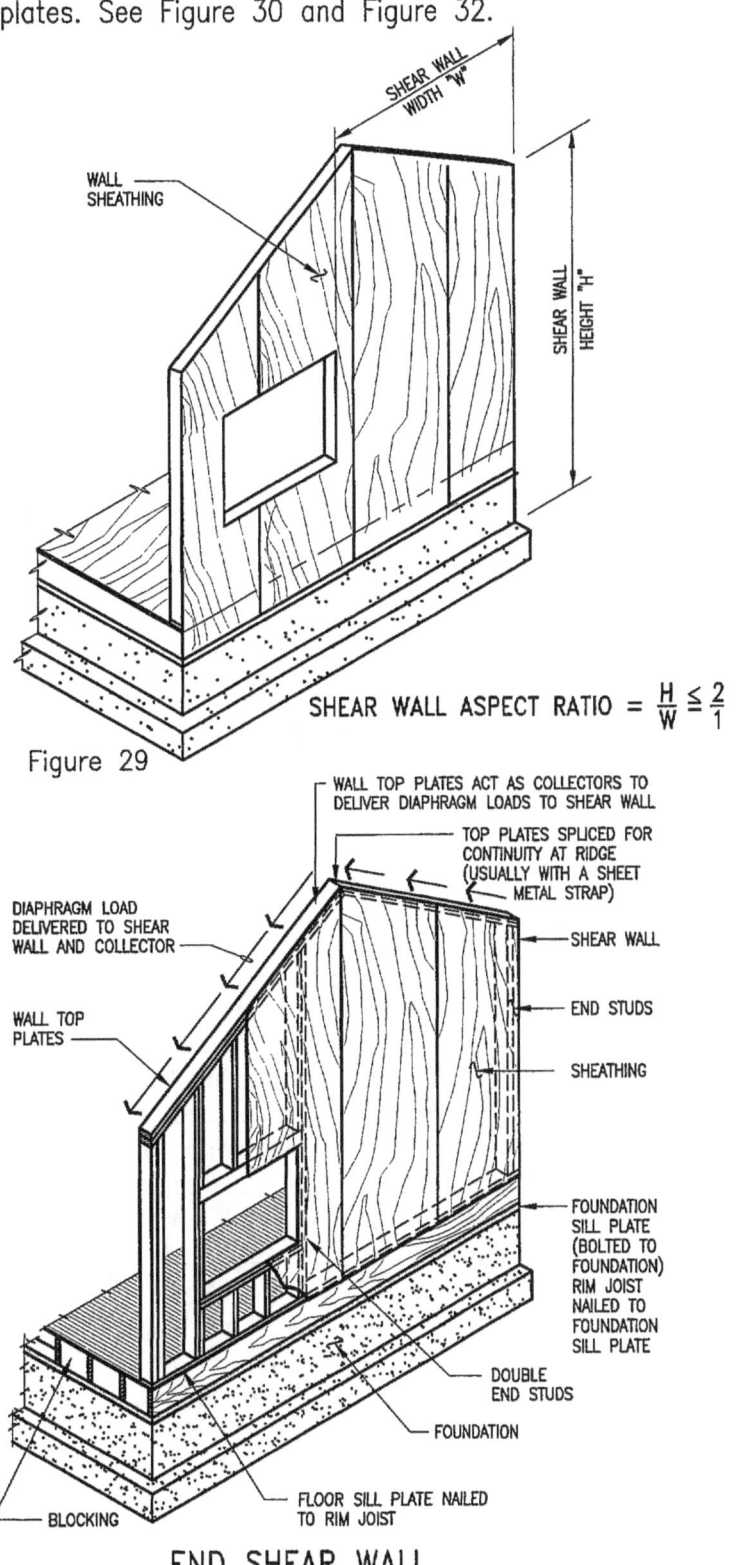

SHEAR WALL ASPECT RATIO = $\frac{H}{W} \leq \frac{2}{1}$

Figure 29

END SHEAR WALL
Figure 30

Wall top plates act as flanges and collectors for the roof and floor diaphragms and can be loaded in tension or compression. Top plates must be spliced for continuity. Wall top plates are stabilized against buckling by nailing the joists or rafters to the plates or by means of a rim joist or end rafter where the framing runs parallel to the wall. Corner and/or end studs act as flanges for shear walls and carry tensile or compressive forces. Only full length studs may be used as flanges and the wall sheathing must be edge nailed to the end studs rather than to adjacent studs. See Figure 31.

Because of the tendency of a tall, narrow shear wall to overturn, a wall with insufficient weight may uplift from the foundation at either end of the wall. To prevent this, the corner or end studs can be anchored to the footing with a hold-down device. See Figure 31 here and Figure 55 on page 29. Experience has shown that hold-down bolts must be inserted in correctly sized holes and must be tight. Bolts should be retightened before closing in the walls. Better performance can be obtained by using 1/4" diameter wood screws to connect the hold-down to the post.

To prevent the building from sliding off of its foundation when exposed to earthquake shaking, the sills must be bolted to the footings. See Figure 33. The footings will then transfer the earthquake loads back to the ground. Bolts should be set accurately on the center line of the sills. A consistent spacing of the bolts should be maintained. Improperly installed sill anchor bolts have contributed to damage in past earthquakes. Care should be taken not to oversize the holes in the sills and the bolts should be tight with a washer under the nut for firm bearing. Use plate washers in high seismic risk areas. Anchor bolts should not be countersunk in the sill. Floor and wall sheathing must be nailed to the framing. See Figures 32 and 33.

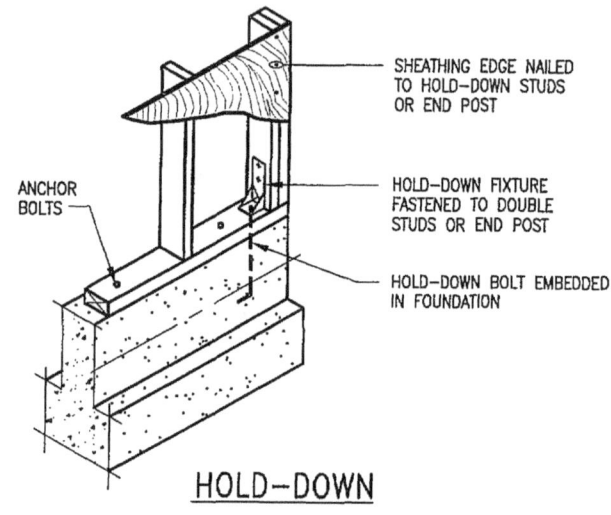

HOLD-DOWN
Figure 31

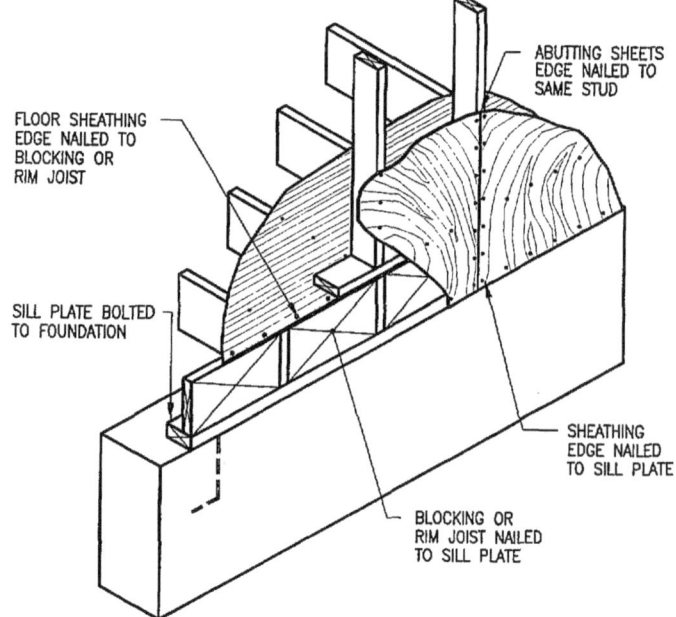

Figure 32

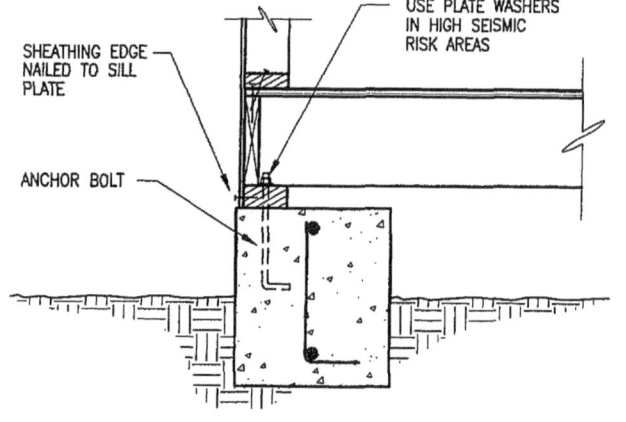

Figure 33

(20)

6. FOUNDATIONS AND FOUNDATION DETAILS

Footing types are dictated by the underlying ground conditions and by their inherent ability to resist forces caused by earthquakes. Some consistent rules apply and should be followed regardless of the footing type selected.

To avoid damage from unequal settlement and subsidence, foundations should be on uniform ground conditions, i.e., footings for the same building should not be part on rock and part on loose fill, soft soil or other significantly different ground conditions unless specially designed to accommodate the non-uniform conditions See Figure 34.

Builders should investigate building sites for ground conditions that could have an unfavorable impact on the foundation. Information noted in Section 4, SITE SELECTION, in the Guide indicates the minimum satisfactory conditions for constructing foundations to resist earthquakes. Local building codes generally provide information regarding minimum standards for foundations. Supplemental information intended for earthquake resistance is contained in the Guide.

Underground basement walls must support the weight of the structure above, floor and roof loads, earth pressures and earthquake loads perpendicular to the plane of the wall. Basement walls acting as shear walls may be thought of as a continuation of the shear wall above and resist the earthquake forces transmitted to these shear walls by the horizontal floor diaphragms, See Figure 35.

Sill plates on basement walls, when properly anchored, keep the shear walls above from sliding or moving. The minimum bolting requirement for moderate and low seismic risk areas is 1/2" diameter by 10" long bolts spaced a maximum of 6'-0" on center with 7" of embedment.

For high seismic risk areas bolts should be 5/8" diameter by 10" long spaced at 4'-0" maximum with 7" of embedment. Bolts should be centered on the sill plate.

Sill plates are normally 2x material, but in areas of high seismic risk, 3x sill plates provide better protection from splitting. Avoid countersunk washers and nuts in sill plates. In areas of high risk, sill bolts should have plate washers under the nuts. Use 3" x 3" x 1/4" plate washers for 1/2" and 5/8" bolts and 3" x 3" x 3/8" plate washers for 3/4" bolts. At ends of sill plates, bolts should be placed 9" from each end of each piece.

Local building codes should be consulted for reinforcing requirements for basement walls. Reinforcement necessary to resist earth pressure and gravity load will usually be adequate for earthquake resistance.

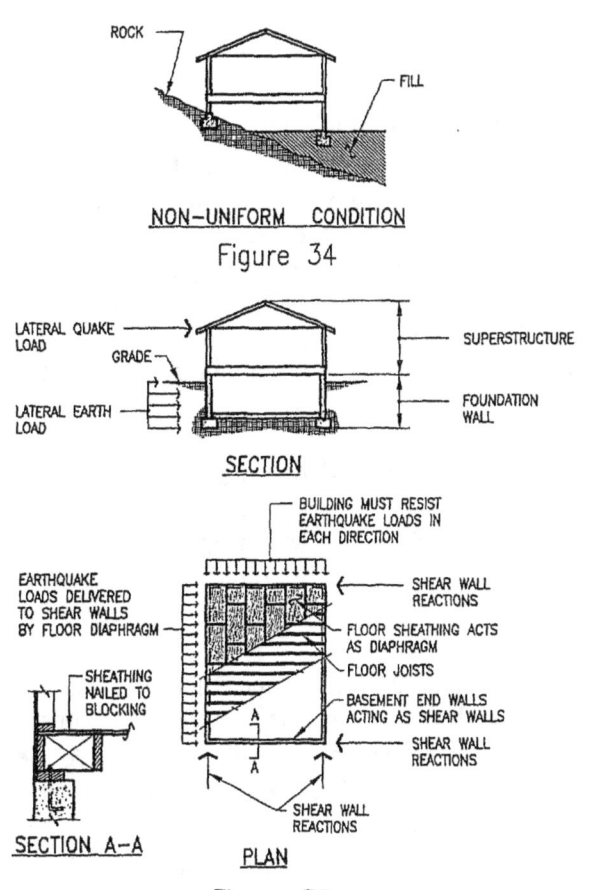

NON-UNIFORM CONDITION
Figure 34

Figure 35

Basement or foundation walls supporting a framed floor where there is a difference in elevation from the top of the walls to the underside of the floor joists may have cripple stud walls supporting the framing. See Figure 36. Cripple stud walls are vulnerable to earthquake shaking and must be braced or sheathed as appropriate for all seismic risk areas. Wall bracing or sheathing for cripple walls should conform to the same requirements as for full height shear walls and can be of any material proposed in this Guide. Plywood or oriented strand board is recommended for high seismic risk areas. Gypsum board sheathing and/or stucco can be used for moderate and low seismic risk areas. Let-in bracing is not recommended for cripple walls except that it may be used in low seismic risk areas.

As observed in the Northridge Earthquake, cripple stud shear walls supported on stepped footings along steeply sloping sites performed poorly. They should not be used in high seismic risk areas. Except for high walls, a concrete stem wall above the footing should continue to the floor framing, thus eliminating the need for a cripple wall.

Continuous perimeter footings are suitable for ground conditions consisting of firm natural sites, engineered landfills and moderate stable slopes. Concrete footings are usually placed in trenches without forms with the bottom of footing at a minimum depth required by code or at least 12" below adjacent grade. Wall sill plates can rest directly on the tops of footings if the footings are high enough relative to the adjacent grade. If a grade beam does not project above the ground, a concrete or masonry stem wall resting on it can be provided. Vent openings and crawl holes may be placed in the stem wall, but the grade beam should be continuous without interruption.

In areas of high seismic risk, footings should have reinforcing. See Figures 37 and 38. Reinforcing is recommended as shown in moderate seismic risk areas, but reinforcing may be omitted in low seismic risk areas if codes and local construction practices permit.

To be effective, reinforcing must be lapped at splices and well tied together at the corners and intersections. Where placed separately, stem walls and grade beams should be doweled together. Sill plates must be bolted to the tops of grade beams or stem walls. See Figure 39. The spacing and size of bolts should be as described for

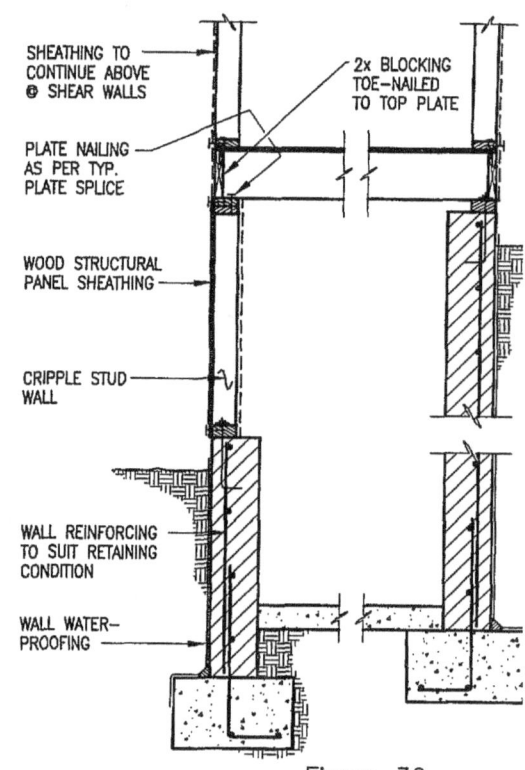

Figure 36

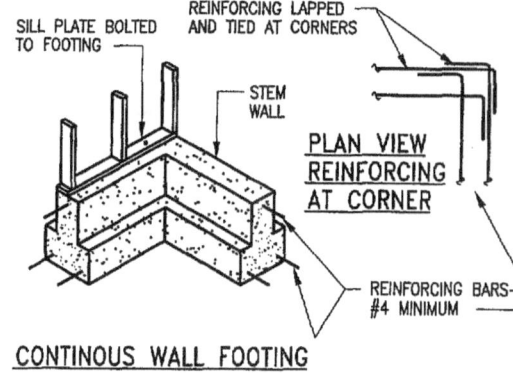

CONTINOUS WALL FOOTING

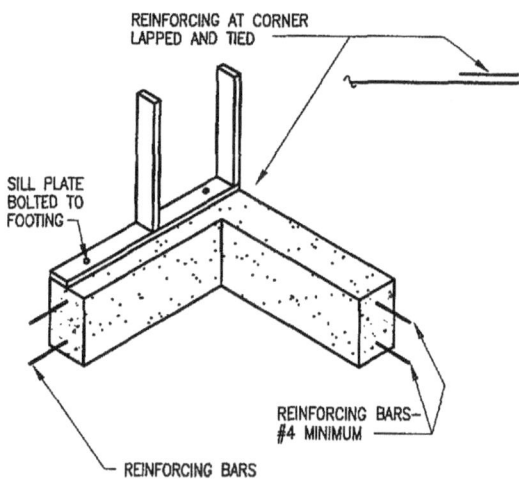

CONTINUOUS WALL FOOTING

Figure 37

basement walls. Hold-down anchor bolts must be embedded in the grade beams as shown in hardware manufacturers' catalogs.

Interior bearing walls or partitions used as shear walls should have continuous footings similar to perimeter foundations. See Figure 38. Reinforcing, where used in footings under interior walls, should be lapped with reinforcing in the perimeter footings at their intersections.

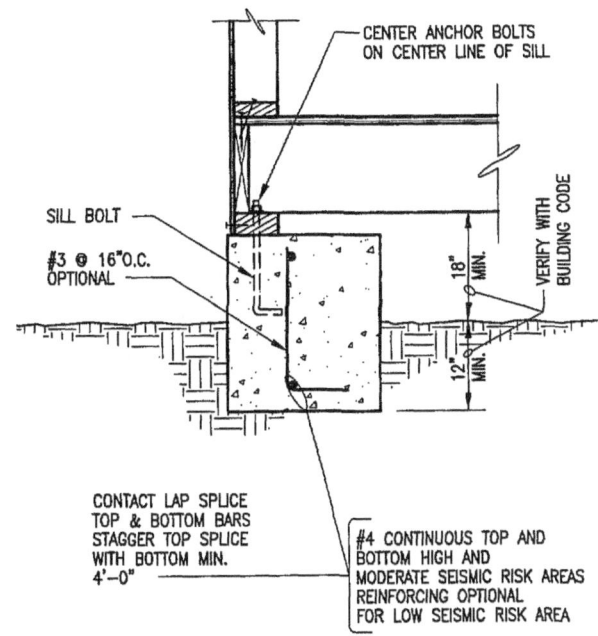

GRADE BEAM

Figure 39

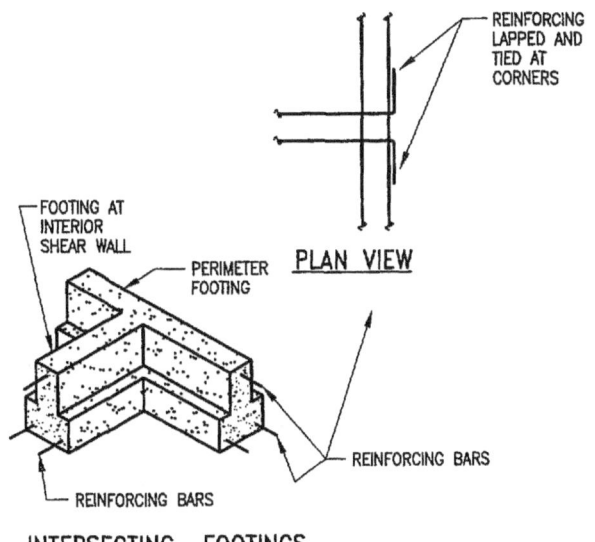

INTERSECTING FOOTINGS

Figure 38

Often, stud walls are prefabricated on the ground and lifted into place. In such cases, bolts holes in the sills should not be oversized to accommodate the preset bolts in the footings. Bolt holes should be marked and drilled in the sill plate using a template made from the known location of the bolts projecting from the footing. After locating and drilling the bolt holes in the sill, the wall can be lifted and set over the anchor bolts.

Floor joists supported on interior bearing lines other than shear walls can be supported on beam and girder framing systems attached to posts bearing on individual spread footings or precast piers. See Figure 40. Sheet metal post caps should be used to connect beams/girders to the posts. If a beam/girder acts as a collector for a shear wall, it must be spliced for continuity with wood cleats or metal straps. Splices should occur only over posts. Girders should be in lengths as long as possible to minimize splices.

and 43. Sills must be bolted to the slab with bolt sizes and spacing as previously described for continuous wall footings.

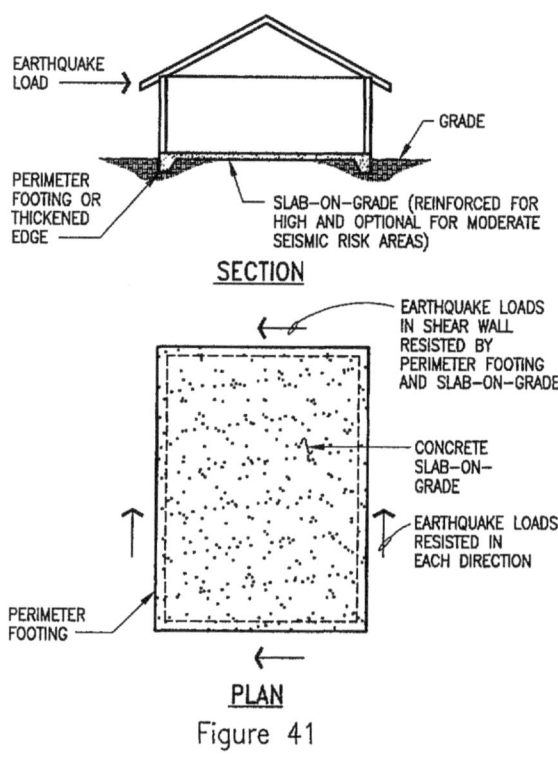

Figure 41

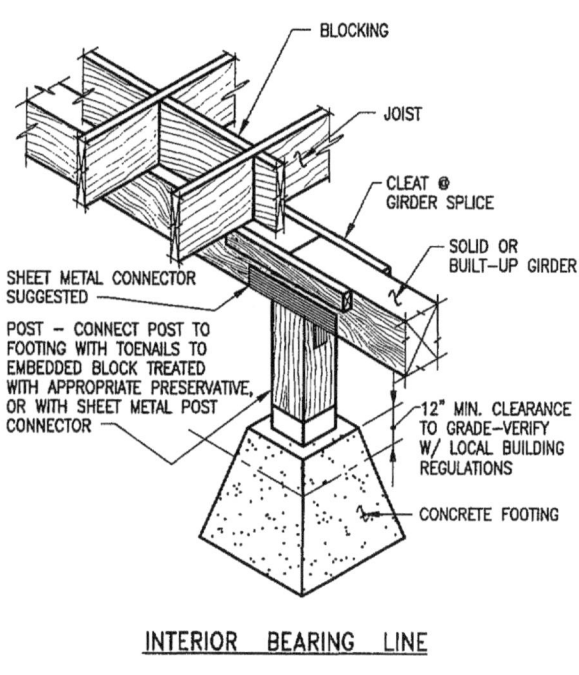

INTERIOR BEARING LINE

Figure 40

Slab-on-grade foundations have performed well in earthquakes. Slab-on-grade construction is a suitable foundation system for most sub-grade conditions. When highly expansive soil or clay soil is encountered, a geotechnical engineer should be consulted. Slabs-on-grade may be used on flat sites and are ideal for homes framed of wood or cold-formed steel with no basement or crawl space. Slabs, unless they are 12" or more in thickness, should have thickened edges at exterior bearing walls and integral thickened trenched footings for interior bearing walls and shear walls. See Figures 41, 42

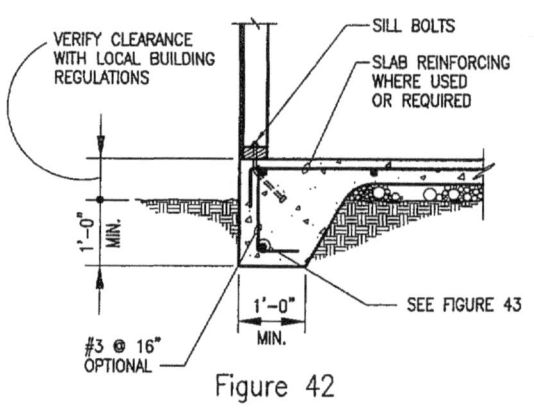

Figure 42

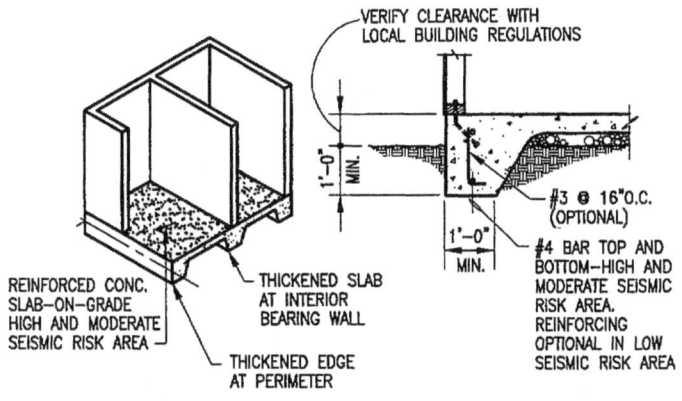

CONCRETE SLAB-ON-GRADE

Figure 43

The slab floor will act as a diaphragm and should have minimum reinforcing in high seismic risk areas so that cracks and differential displacements do not impair the diaphragm properties. Slab reinforcing is also effective in bridging across soft spots and depressions in the subgrade regardless of need for seismic resistance. Reinforcing is optional in low seismic risk areas. At a minimum, welded wire fabric is recommended to keep slabs integral and minimize cracking.

Drilled-in-place concrete-filled piers with grade beams spanning between the piers are suitable for compressible soils, loose land fills, and known weak soil conditions. Sizes and depths of piers must be determined from local building regulations or from a geotechnical consultant. Reinforcing of the grade beams must be provided to support the weight of the building and contents. Reinforcing required for normal loading will be adequate for earthquake loads as well. Reinforcing should employ individual bar lengths as long as practical and be lapped and tied at splices and at corners and intersections. The piers should be doweled to the grade beams. See Figure 44.

Some foundation types are not suitable for seismic resistance without special consideration related to either the site or special requirements of the building.

See Table No. 2 on page 26. Four conditions not covered in the Guide are listed below:

1. Pole houses.

2. Houses on pile supported platforms.

3. Floating foundations on unengineered hydraulically placed landfill.

4. Isolated footings not connected by continuous grade beams

Information and details of construction shown in the Guide may not be universally suited to these foundation types. Construction should not be attempted without the assistance of a design professional.

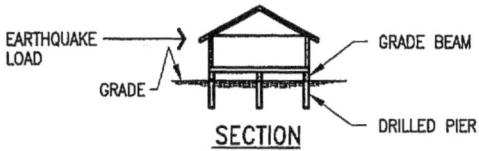

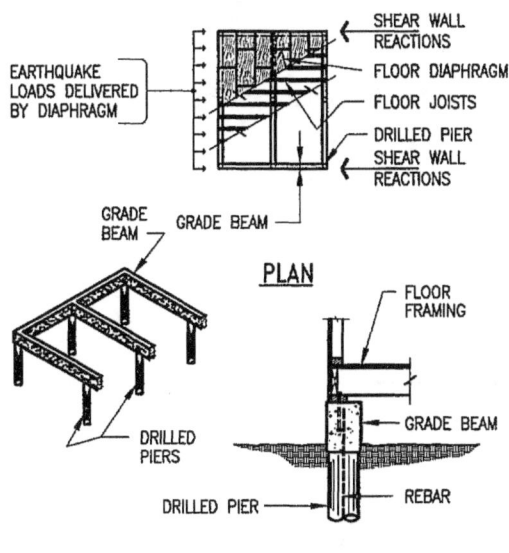

Figure 44

(25)

Table No. 2 – Foundation Types Requiring Special Construction

Pole House	
Pile Supported Platform	
Individual Isolated Piers	
Floating Foundation on an Unengineered Landfill	

7. FLOORS

Floor diaphragms are significant structural elements especially in two story construction or if a first floor is required to span horizontally across a basement or over a crawl space. Ideally, the floor should be regular in shape and all in the same plane without vertical offsets. A simple rectangular, symmetrical shape (box) is the best configuration. See Figure 45. Where possible the earthquake forces in the floor diaphragm should be resisted by a symmetrical pattern of shear walls supporting the boundary of the floor.

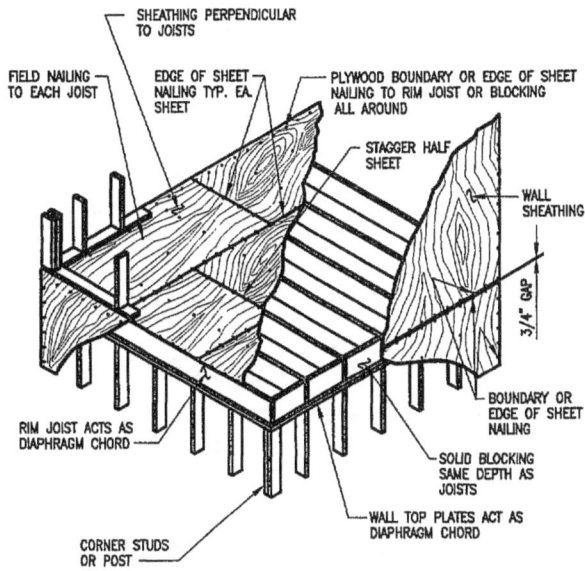

TYPICAL DIAPHRAGM
Figure 46

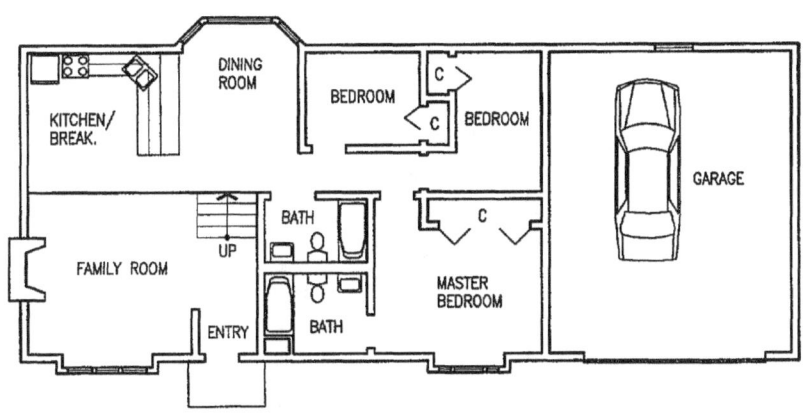

Figure 45

A diaphragm consists of joists, sheathing and edge members called chords. See Figure 46. Sheathing on floors serves the dual purpose of supporting the load of furnishings and occupants and of transferring earthquake or wind loads to the shear walls. Commonly used sheathing materials, when adequately nailed, carry the vertical floor loads and resist the earthquake loads in the plane of the diaphragm. Nailing for diaphragm action is generally greater than the nailing required for gravity loads.

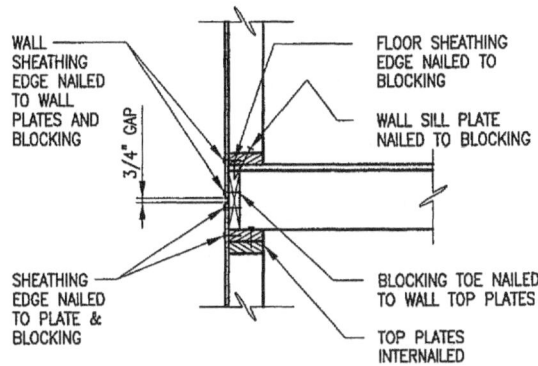

WALL PERPENDICULAR TO JOISTS
Figure 47

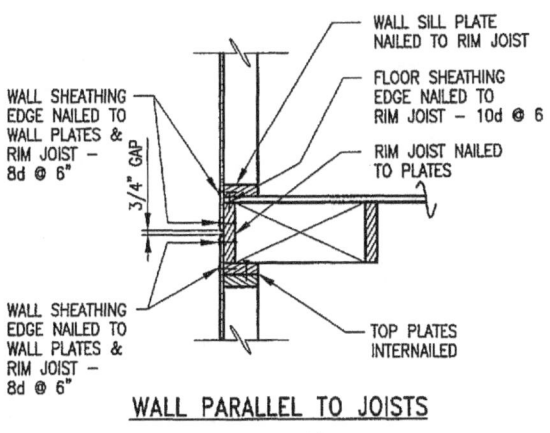

WALL PARALLEL TO JOISTS
Figure 48

Plywood or oriented strand board for subfloors may be used in any seismic risk area. Tongue and groove sheets should be oriented with the long edge perpendicular to the framing and need not be blocked. The short edge of abutting sheets must be edge nailed to the same framing member. To achieve a stiffer diaphragm in areas of high seismic risk, square edge sheets with the long edges of each sheet nailed to blocking and short edges nailed to a common framing member should be used. Straight sheathing without a Wood Structural Panel overlay should be avoided except in low seismic risk areas.

For two story construction or floor framing supported on stud walls, top plates act as flanges or chords for floor diaphragms. They should not be interrupted by openings such as windows, ducts, piping or conduits, unless chord continuity can be satisfactorily restored with straps or ties. Top plates must be internailed and receive additional nailing at lap splices. See Figure 47, 48 and 49. Wall top plates must continue across the full width and length of the floor diaphragm from corner to corner without interruption. Diaphragm sheathing must be edge nailed to blocking or rim joists that are toe nailed to the top plates.

To function properly diaphragms must conform to specific shape limitations that are controlled by the diaphragm span-to-width ratio discussed before. If the allowable ratios are exceeded it is necessary to provide additional shear walls to divide the diaphragm in sections so that the sections meet the prescribed span-width diaphragm limits. Shear walls (panels) in line at the edge of a diaphragm, but separated, can be tied together by the wall top plates acting as collectors or drag struts between the shear walls. The top plates must be spliced for continuity. See Figure 50. Shear walls not at the edge of a diaphragm or within the diaphragm borders can also be used by employing light gage strap ties nailed to the shear wall and extending into and nailed to the diaphragm framing. See Figure 51.

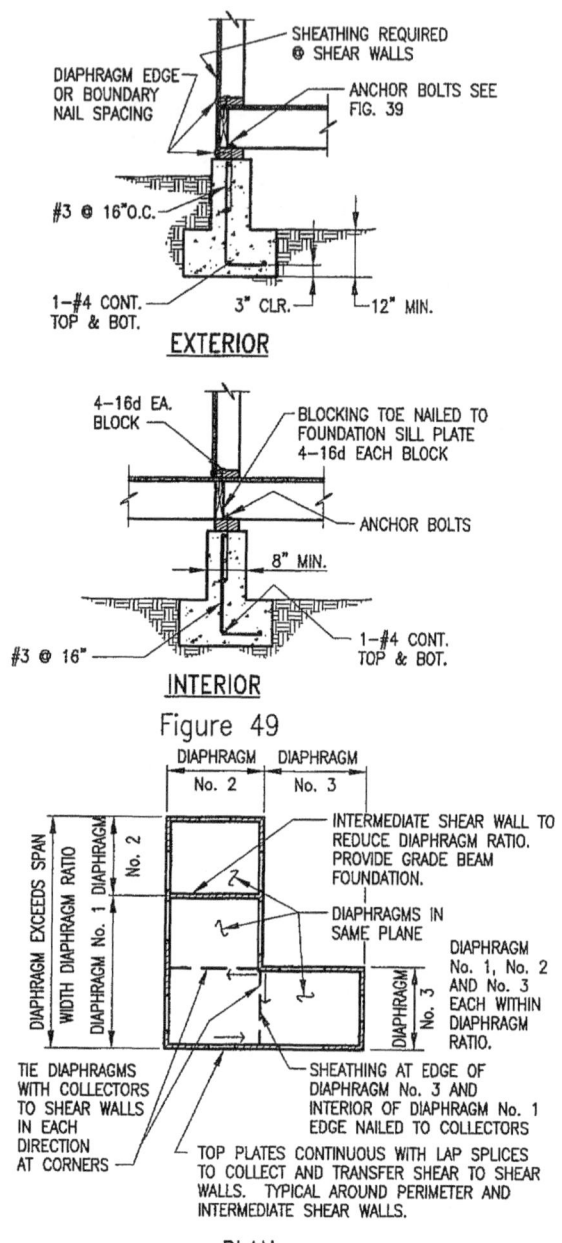

Figure 49

Figure 50

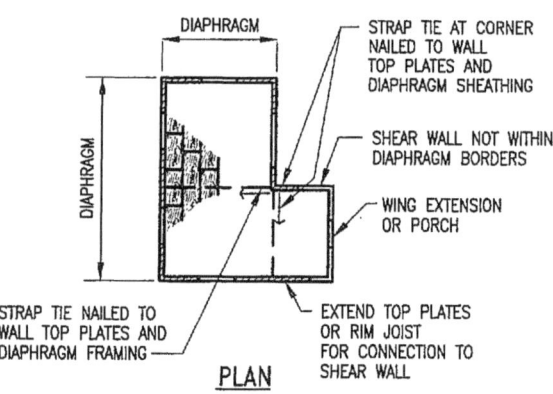

Figure 51

(28)

As for building wings, collectors or ties can be used to transfer earthquake loads developed in areas of a floor not supported on walls such as at porch enclosures that are open underneath and supported on posts. Collectors must be nailed to floor framing within enclosures and extend into and be nailed to floor diaphragm framing as shown in Figure 51 and 54.

Sill plates must be bolted to the top of the footing or stem wall. See Figures 49 and 53. In high seismic risk areas, chances of plate failure are reduced by using a larger number of anchor bolts over the width of the shear wall than may be prescribed by Code and by the use of plate washers. Hold-down anchor bolts, where needed, must be embedded in the footing as shown in manufacturers' catalogs. See Figures 31 and 55.

Where collectors or drag struts continue across a break in diaphragm planes the collectors or struts must be anchored to the vertical framing to prevent uplift when they are loaded by earthquake forces. See Figure 52.

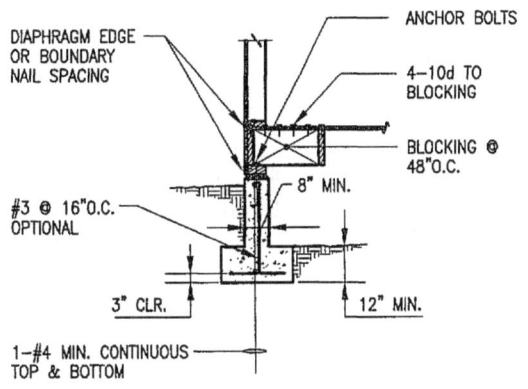

SHEAR WALL SECTION
HORIZONTAL FRAMING PARALLEL TO WALL
Figure 53

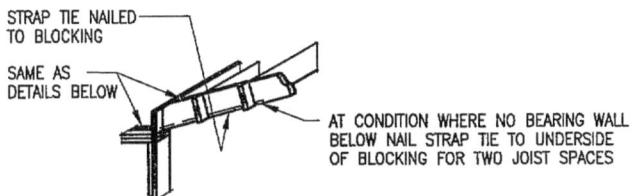

Figure 54

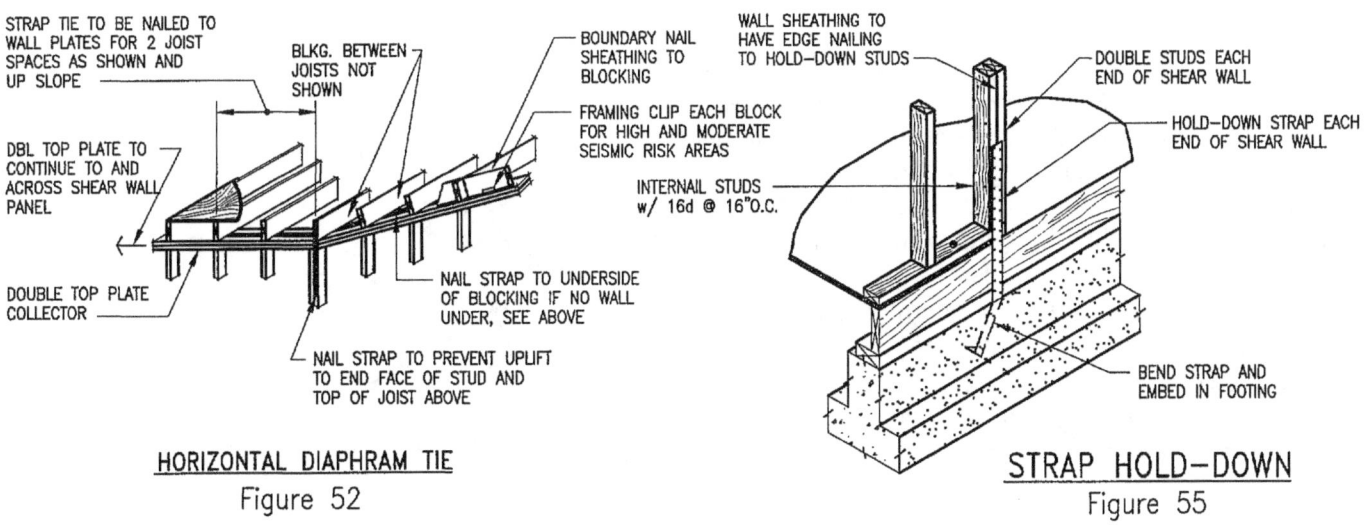

HORIZONTAL DIAPHRAM TIE
Figure 52

STRAP HOLD-DOWN
Figure 55

(29)

A floor diaphragm or any of its parts should, where possible, be in one horizontal plane between shear walls and not be broken by vertical offsets or ramps. Where a floor plane is interrupted, as in split level construction (see Figure 56), the floor will then be composed of more than one diaphragm. Each of these diaphragms must have a boundary shear wall along each edge. At the offset edge, each diaphragm may use the common shear wall to resist the earthquake forces in each diaphragm. See Figures 58 and 59 for split level floor framing details.

Openings in floor diaphragms must be limited. Penetrations at corners should be avoided. Small openings for furnace ducts and registers, flue stacks and chimneys are permissible. See Figure 57. Large openings for stair wells must receive special framing and nailing of floor sheathing around the opening. Where any opening dimension reduces the effective length or width of a diaphragm by 50% or more, a shear wall should be placed parallel with the narrow edge of the diaphragm on each edge of, or in line with, the edge of the opening.

Ideally, floor diaphragms should be fastened with screw shank nails with a diameter equivalent to common or box nails. If nails other than common wire nails, such as box, sinker or staples are used, the spacing should be reduced from those prescribed in codes for common nails. To prevent squeaking a bead of construction adhesive should be applied on the top edge of floor joists when laying the sheathing. The use of adhesive in this manner has been shown to increase stiffness and improve performance of floor diaphragms in earthquakes.

In addition, nominal nailing (at 10" to 12" spacing) must be provided along framing members located between the edges of the sheets. This nominal field nailing secures the sheathing to the framing and prevents buckling of the sheets when the floor functions as a diaphragm during an earthquake. When using a nailing gun, care must be exercised to avoid overdriving the nails. The gun operator should also make sure that the nails penetrate the framing members below. It is good practice to locate the members below by snapping a chalk line on the top face of the sheathing. A new nail should replace each nail not properly driven.

Blocking or rim joists should be nailed to top plates with 4–10d toe nails each block or 10d toe nails at 4" centers for rim joists. Light gage angle clips can be used in lieu of toe nailing.

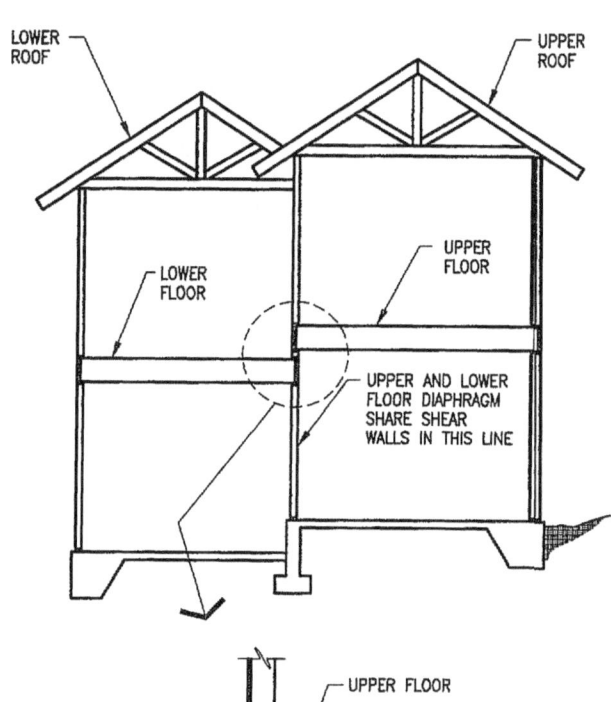

Figure 56

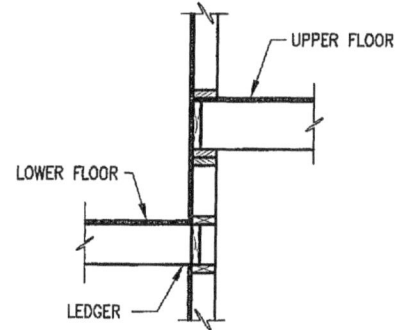

Figure 57

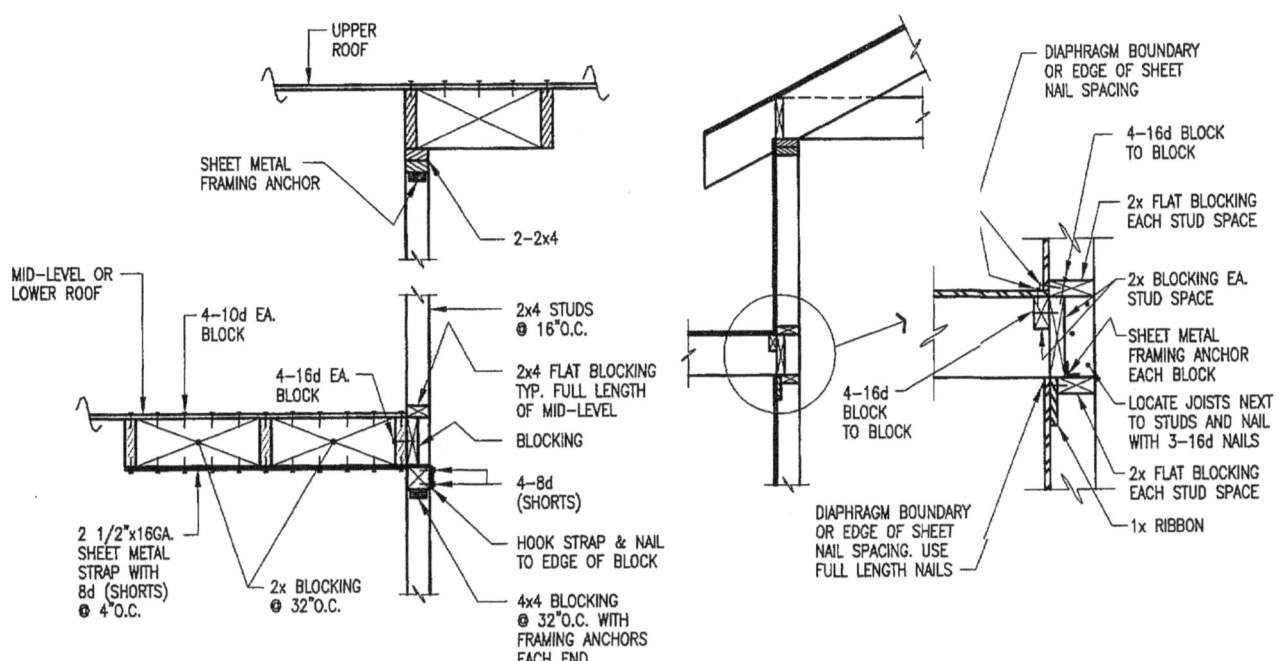

SPLIT-LEVEL TIE
FLOOR FRAMING PARALLEL TO WALL

Figure 58

SPLIT-LEVEL TIE @ 32" o.c.
FLOOR FRAMING PERPENDICULAR TO WALL

Figure 59

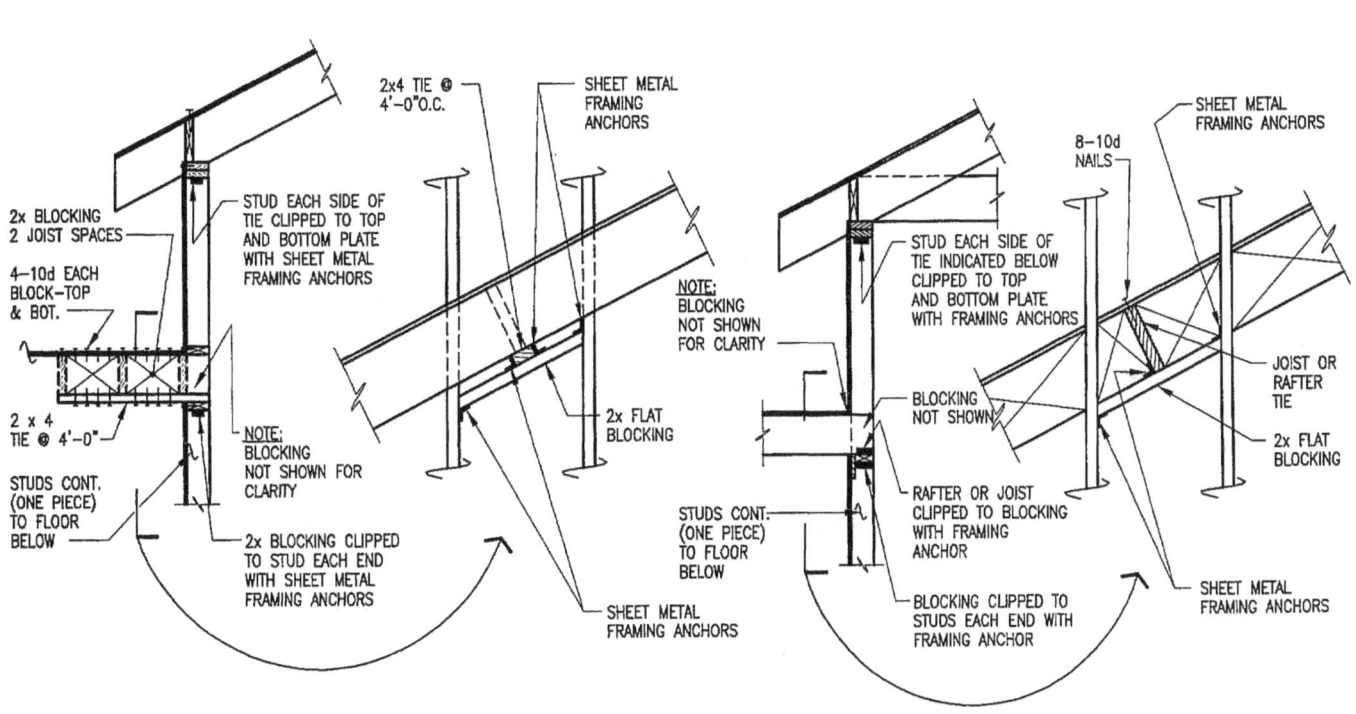

SPLIT-LEVEL TIE
SLOPING ROOF FRAMING PARALLEL TO WALL

Figure 60

SPLIT-LEVEL TIE @ 32" o.c.
SLOPING ROOF FRAMING PERPENDICULAR TO WALL

Figure 61

8. SHEAR WALLS

Shear walls resist earthquake loads transmitted to them by floor and roof diaphragms. A shear wall is required along each side of the perimeter of the floor or roof (horizontal) diaphragm, usually in each exterior elevation. The most effective locations for the exterior shear walls are at the corners of the building where the walls are mutually perpendicular to each other and have a common point of intersection. See Figure 62. To maintain symmetry, the same length of wall should be provided at each of parallel exterior walls. In order to minimize torsional rotation which can increase earthquake loading on shear walls, an unbalanced condition should be avoided. See Figure 63. Some codes limit the maximum distance between shear walls and specify the percentage of an exterior wall that must be sheathed. For long walls, this may require placing one or more shear panels between the corners. See Figure 64 below and Figure 67 on page 34. If shear walls can not be placed at the corners, they should be located as near to the corner as possible.

Shear wall configurations should conform to ratios that limit height to width. See Figure 29 on page 19 and Table 3 on page 39. With a ratio of 2:1 for height to width, and considering a usual story height of 8'-0", the minimum width of wall to be provided at each corner of each exterior wall should be 4'-0" for moderate and high seismic risk areas. This ratio limit is also recommended for low seismic risk areas.

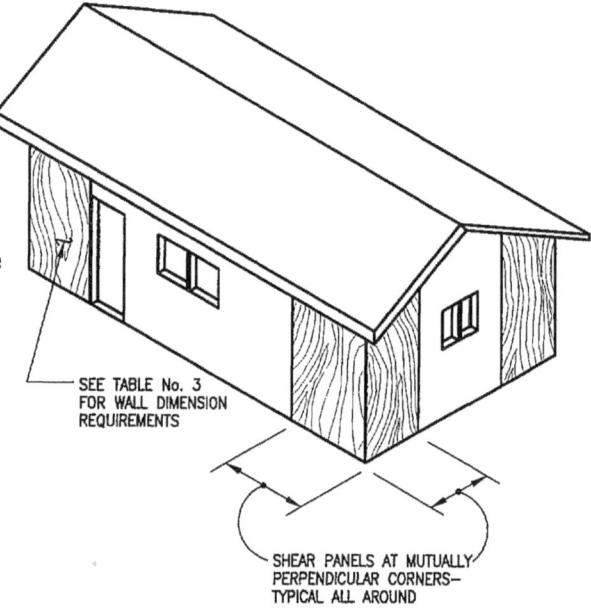

Figure 62

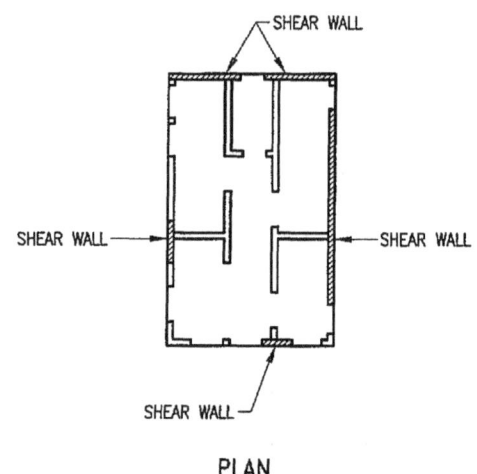

UNBALANCED SHEAR WALL CONDITION

Figure 63

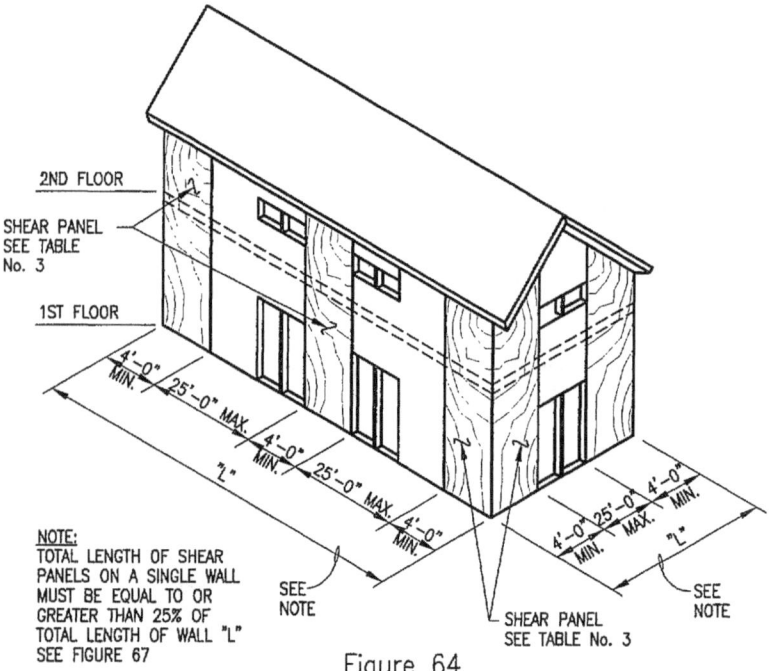

Figure 64

(32)

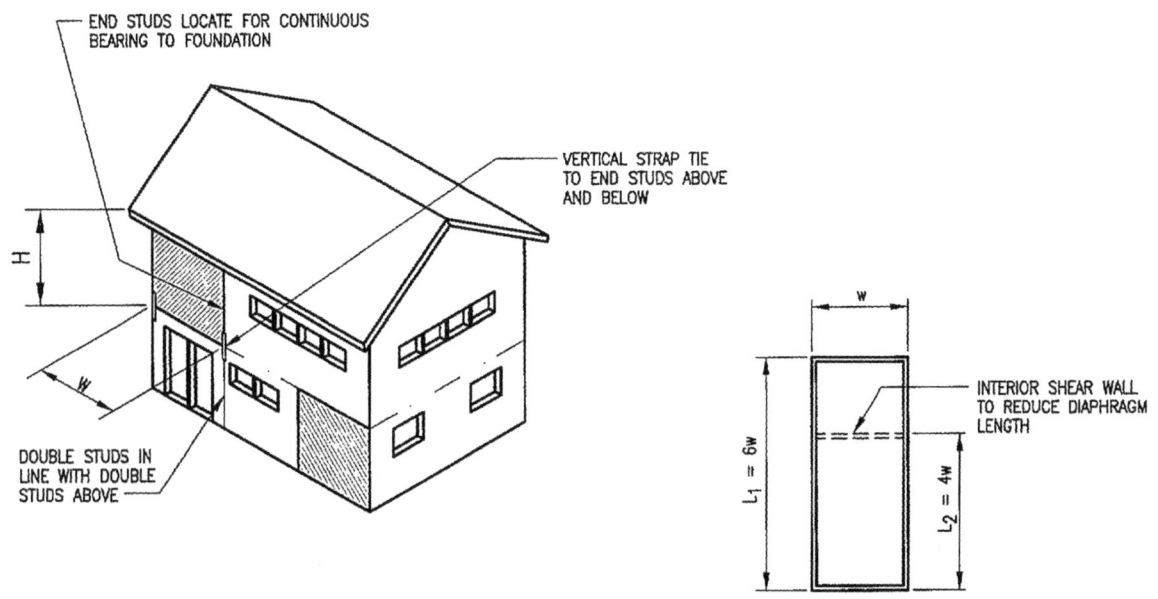

OFFSET SHEAR WALL

Figure 65

PLAN

Figure 66

Where horizontal diaphragm ratios are exceeded, interior shear walls are needed to reduce the diaphragm ratio. See Figure 66. Because interior shear walls are permanent and may interfere with flexibility in modifying floor layouts, care should be taken in determining their locations. When needed, interior shear walls should be walls that are natural separations such as garage walls or other walls regarded as permanent and should have a minimum number of openings. Plumbing walls should be avoided for use as shear walls.

The reasons for avoiding torsional rotation of buildings has been previously discussed. In equalizing the length of shear walls on all sides of a building, it is important to remember that the difference in stiffness in long and short walls is not directly proportional to the lengths of the walls. Equivalent stiffness of shear walls cannot be obtained by providing two separate 8' lengths of wall along one side and one 16' length along the other. Shear walls should not be spaced more than a prescribed distance apart (varies by seismic risk area, see Section 14 BUILDING CODES). Intermediate shear panels will be needed in long exterior wall lines.

Shear walls should extend continuously from foundation to roof line in single-story buildings and from foundation to floor line and floor line to roof line in two-story buildings. In two-story construction, it is not necessary for the shear walls to line up one above the other, but such continuity is encouraged to increase the stiffness of the wall. See Figures 65 and 68. If the walls in the top story are offset from the walls below, the overturning forces in the end studs or posts of the wall above must be carried down through the wall framing below to the foundation. The end studs could be located above a header, but special precautions must be taken to secure the studs to the header and the header to its supporting framing. Cripple studs should be placed under each end of the header. The studs supporting the header should be tied to the sill plate with framing clips or a hold-down.

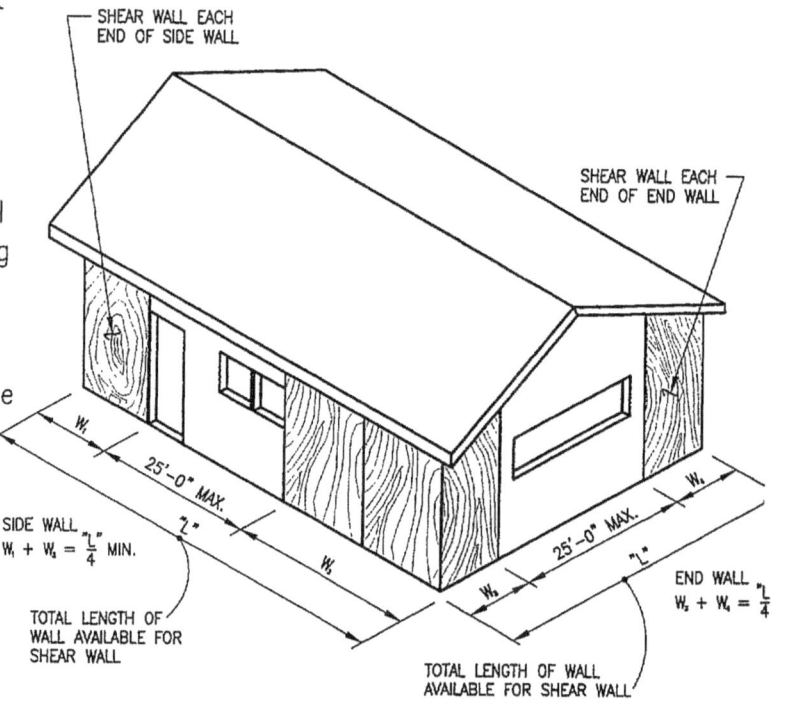

Figure 67

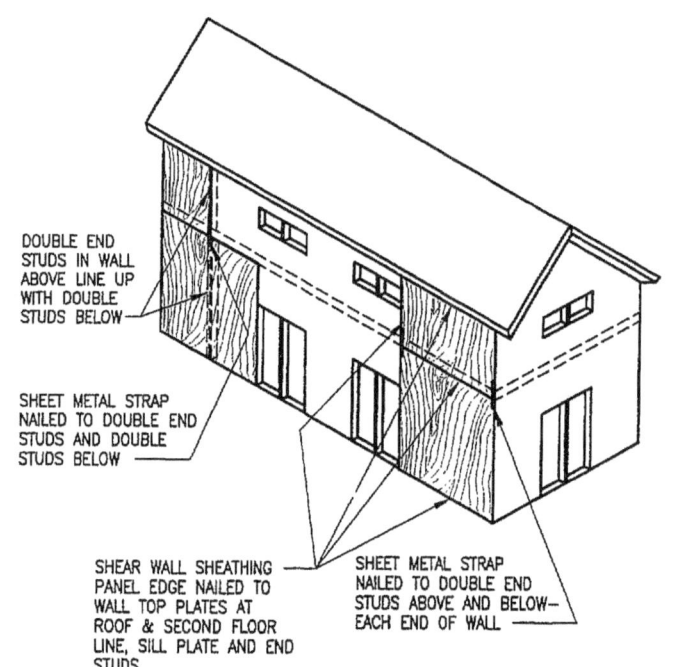

SHEAR WALL 2 STORY
HIGH SEISMIC RISK AREA

Figure 68

(34)

Shear walls should be solidly sheathed with limited openings for doors or windows. Small penetrations for vent openings are permissible, but there should not be more than one opening for every eight feet of wall length. Penetrations or openings should be avoided in 4'-0" long shear walls. Doors and windows should be centered in a wall, where possible; but, in any case, the edge of any opening should not be positioned within sixteen inches (16") of the vertical edge of a shear wall. The wall should extend a minimum of two feet eight inches (2'-8") beyond the other vertical edge of an opening. Wherever possible, the wall sheathing should continue above and below an opening. See Figure 69. In no case should openings interrupt sill plates or wall top plates.

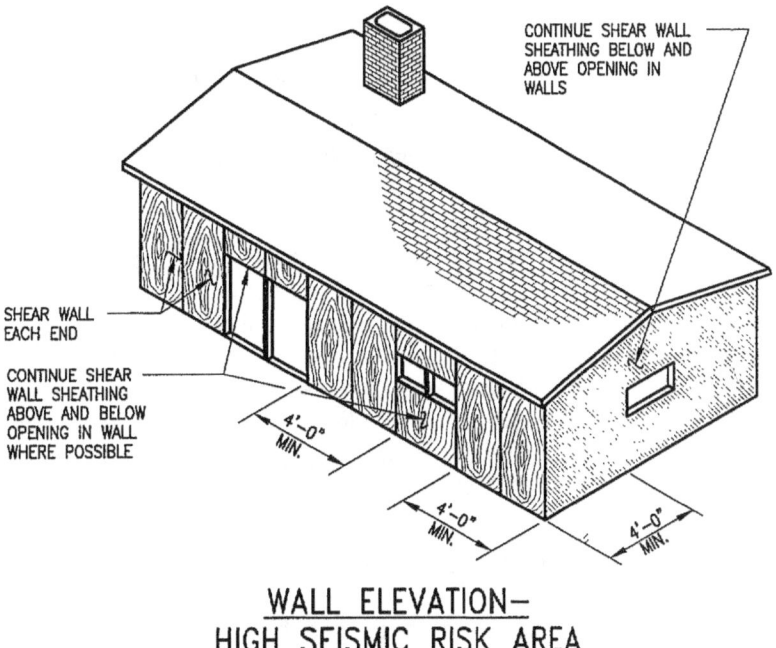

WALL ELEVATION— HIGH SEISMIC RISK AREA

Figure 69

Cripple stud walls are sometimes used to support floor framing over crawl spaces, partial height basement walls, stepped foundations, and split-level floor framing. Cripple walls must be sheathed as shear walls to transfer earthquake forces from the walls above the foundations. The widths of cripple shear walls and the construction details should be the same as for shear walls on the floor above, although the individual walls can be offset. In no case should there be less width of shear wall available in the cripple wall than there is in the wall above.

Four types of sheathing are permitted by code for shear walls. See Table No. 3 on page 39. Ideally, the same type of sheathing should be used throughout any level of any building. Sheathing should conform to the prescriptive requirements of applicable building codes.

1) Wood Structural Panel (plywood or oriented strand board) for exterior and interior shear walls should be a structural grade with exterior glue and a minimum thickness of 3/8". When installed with long edges vertical, sheets should have vertical butt joints on the centerline of the same stud and the sheathing should extend from the sill plate to the wall top plates, rim joist or blocking. See Figures 70 and 71. Sheets may be installed with long edges horizontal, but the horizontal edges of each sheet should be nailed to flat blocking between studs. Sheathing should be edge nailed along all edges of each sheet. Wider spaced nailing to intermediate studs in the field of each sheet is also required. For high seismic risk areas, a 5 ply, 1/2" thick structural grade plywood is recommended for best performance and for added stiffness to reduce damage. See Figures 70, 71, 75 and 77.

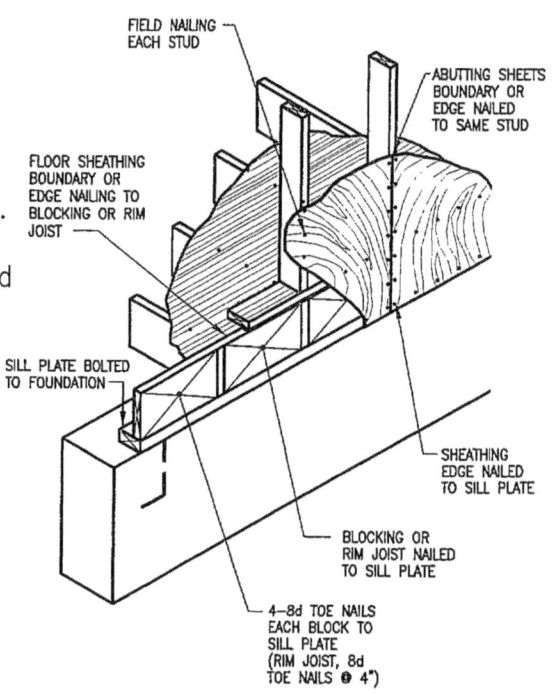

Figure 70

2) Diagonal sheathing, although not commonly used, is still permitted by code in all seismic risk areas. If used it should consist of 1" nominal boards installed at an angle of 45 degrees to the studs. The boards should be nailed to each stud with two 8d nails for 1x 6 boards and three 8d nails for 8" wide boards. In addition, 6" boards must be edge nailed to sill plates, wall top plates and wall end studs with three 8d nails and 8" boards must be nailed with four 8d nails. Butt joints of adjacent boards must be separated by at least two stud spaces.

3) Gypsum board may be used in all seismic risk areas, but experience from the Northridge Earthquake indicates that gypsum board should not be used in high seismic risk areas unless significant damage in the event of a strong earthquake is considered acceptable. The wall board may be applied with the long dimension parallel or perpendicular to the studs with flat blocking provided for all unsupported edges. Although gypsum board is discouraged in areas of high seismic risk, it is considered adequate as sheathing for shear walls in moderate and low seismic risk areas. See Table No. 3. For high and moderate seismic risk areas gypsum wall board sheathing should only be used on very long walls in order to keep stresses low. It is recommended that at least 50% of any exterior wall should be sheathed and the minimum length of each shear panel should be 8'-0".

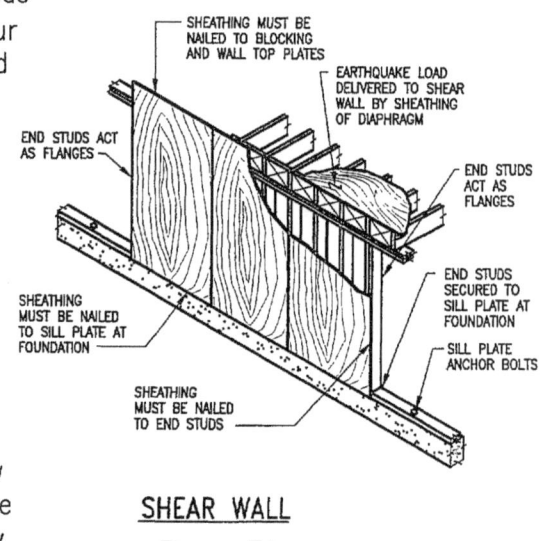

SHEAR WALL

Figure 71

4) Stucco applied over metal lath nailed directly to the studs can also be used to resist earthquakes. In high seismic risk areas, self-furring lath may not be reliable.
Lath should be nailed to the studs, sills and wall top plates. Applying the lath with staples is not recommended in high seismic risk areas. In the Northridge Earthquake, it was noted that where stucco was used with lath or furring applied directly to the studs extensive damage resulted from the earthquake shaking. Where stucco is used in high seismic risk areas, it is recommended that plywood or oriented strand board be used on shear walls with the lath applied to the face of sheathing. Metal lath can be nailed directly to the face of studs on portions of the walls not sheathed.

In high seismic risk areas, the overall performance of a building can be improved by continuing the sheathing across the entire wall from end to end. The sheathing should continue above and below window and door openings. See Figure 69. The type of sheathing should be uniform throughout. Plywood or oriented strand board is recommended.

Proper nailing of the sheathing is very important. Only nail types specified in building codes should be used and substitution of sinker or cooler nails for box or common nails is not permitted. Nailing guns should be adjusted to prevent overdriving and damaging the sheathing. Edge distances and nail spacing should be maintained. In high seismic risk areas, staples should not be substituted for nails unless staples have been tested for use with the specific application.

Two methods are available for bracing steel stud shear walls. Conventional practice is to use light gage steel diagonal strap bracing. See Figures 78 and 80 for wall bracing and Figure 79 for bracing at garage door openings. The straps should be in an "X" configuration or diagonally opposing each other with the straps being one piece from the sill plate to the wall top plates. Straps are secured to light gage steel gusset plates with sheet metal screws The straps should be attached to the studs at each stud crossing. Plywood sheathing can also be used over steel stud framing. Screws must be used to fasten the plywood to the studs and plates. Spacing and edge distances should be the same as for nails.

In the discussion of seismic resistant systems, it was noted that shear walls may tend to lift up at corners when resisting earthquake loads. To prevent this, the end studs or posts of shear walls should be anchored to the foundation with hold-down devices. See Figure 72. Hold-down anchors are usually needed at ends of shear walls where walls run parallel with the joists or rafters. Where bearing loads on the walls are sufficient to keep the walls from lifting at their ends hold-downs are not required.

Figure 72

(37)

Several types of hold-down anchors are available and the types used should be based on the amount of uplift expected. i.e. short shear walls in high seismic risk areas need stronger devices. The most effective hold-down devices are either bolted or screwed to posts and bolted to the foundations. Where used they should be provided at each end of a shear wall. Hold-down bolts should be retightened just before closing in the walls. Other hold-down devices are available that can be embedded in the foundation and nailed to the end studs or posts. Strap hold-downs are desirable for all shear walls but can be omitted in areas of low seismic risk or as noted above. See Figure 76.

An attached garage on the side or end of a house may present a special problem as narrow wall elements on each side of the garage door opening may not function effectively as shear walls and will not, in most cases, meet height to width ratio requirements. To prevent rotation of the garage roof diaphragm special construction of the walls perpendicular to the garage door opening is required or a rigid steel frame can be put across the open front. See APPENDIX pages (A4) and (A5). If a shear wall is available adjacent to or in line with the garage door opening, it may be used to stabilize the garage and must be constructed as a conventional shear wall with earthquake loads transferred to the wall by continuous top plates or other collector or drag strut members. See Figure 73.

Where walls parallel to the roof framing do not extend uninterrupted to the roof sheathing they may require lateral support at the level of the top plates. See Figure 74. This detail is applicable primarily in high seismic risk areas, but should be considered for all seismic risk areas where wind loads are a concern. For gabled roofs, where this condition is most likely to occur, it is better to use full length studs from foundation to the roof plate line.

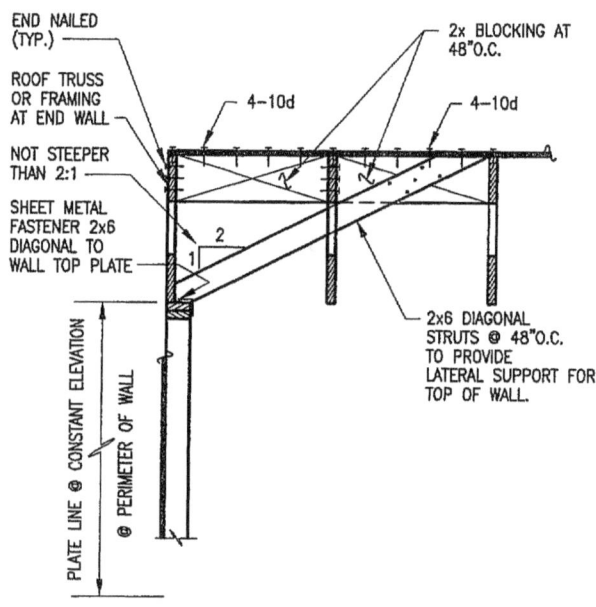

Figure 74

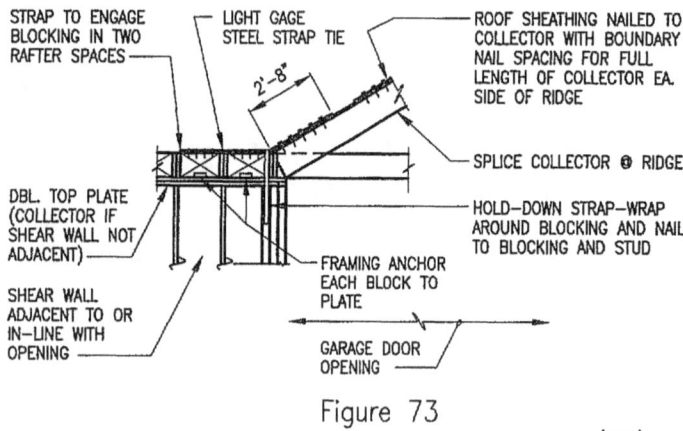

Figure 73

Table No. 3 Maximum Floor and Roof Diaphragm and Wall Aspect Ratios

	Sheathing		Seismic Risk	L/W	H/W
ROOF DIAPHRAGM	Wood Structural Panel	Plywood	High	4*	
			Moderate	4*	
			Low	4	
		Oriented Strand Board	High	4*	
			Moderate	4*	
			Low	4	
	Diagonal Sheathing		High	2	
			Moderate	2	
			Low	2	
	Straight Sheathing		Low	2	
FLOOR DIAPHRAGM	Wood Structural Panel	Plywood	High	4*	
			Moderate	4*	
			Low	4	
		Oriented Strand Board	High	4*	
			Moderate	4*	
			Low	4	
	Diagonal Sheathing		High	2	
			Moderate	3	
	Straight Sheathing		Low	2	
SHEAR WALL	Wood Structural Panel Plywood Oriented Strand Board		High		2
			Moderate		2
			Low		$3\frac{1}{2}$
	Gypsum Board Stucco		High		2
			Moderate		2
			Low		3
	Diagonal Sheathing		Moderate		2
			Low		$3\frac{1}{2}$

* All edges blocked. Reduce to 3 for unblocked edges.

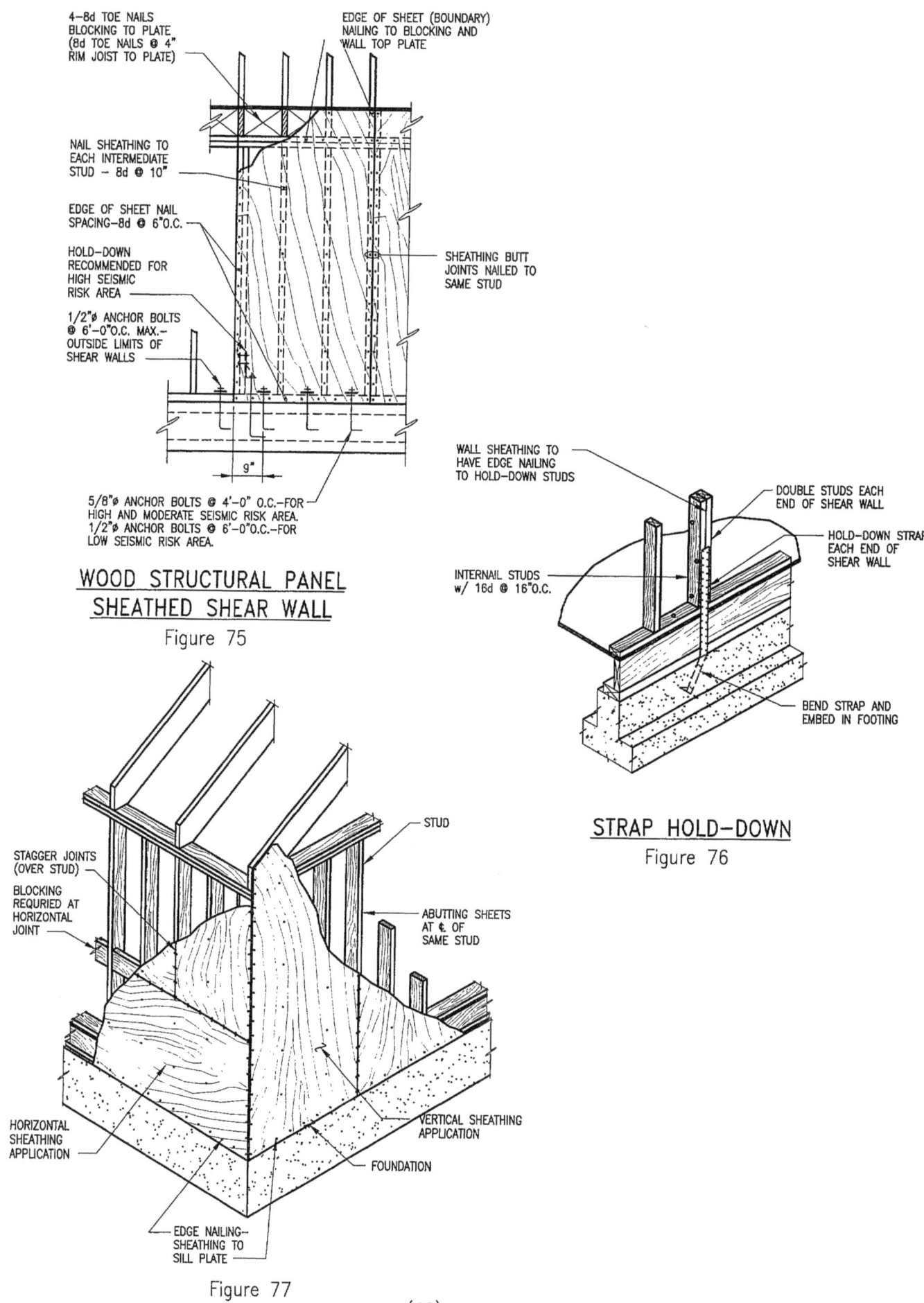

Figure 78

Figure 79

Figure 80

9. ROOFS

Roofs act as horizontal diaphragms and are essential elements in transmitting earthquake induced horizontal forces to the shear walls. Although gypsum board finish materials on the underside of the ceiling joists will assist the roof sheathing in transferring horizontal forces, it is best to assume (except in areas of low seismic risk) that the roof sheathing acts alone in resisting the earthquake. Roofs must conform to certain diaphragm ratios or plan dimensions and shapes. Variations in form are permissible provided the perimeters are supported by shear walls and continuity is maintained at openings in the diaphragm. The perimeter or boundary edges of roof diaphragms should be nailed to blocking or rim joists that are toe nailed to wall top plates or collectors so that the earthquake forces from the diaphragms can be resisted by the shear walls. See Figure 81. Openings in diaphragms are permitted, but should be limited in size and location.

Simple, regular shapes symmetrical about each axis with unbroken planes provide good diaphragm performance. A flat or slightly pitched roof with square or nearly square configuration is best for good behavior as long as the roof plane is not interrupted by large openings or rendered discontinuous at the boundaries by cut-outs or notches that destroy continuity or symmetry of the diaphragm.

It is possible to provide some variety in form when using square, rectangular or other symmetrical shapes. This is especially true of flat roofs that can be sectionalized and tilted at different angles or by having one section dropped into a different plane. Consider that diaphragms in different planes become separate diaphragms and each diaphragm must meet all requirements independently. See Figure 82.

Roofs at different levels are similar to floors with split level configurations and require special connection details at common walls. See discussion of split level floors on page 30 and roof details Figures 60 and 61 on page 31.

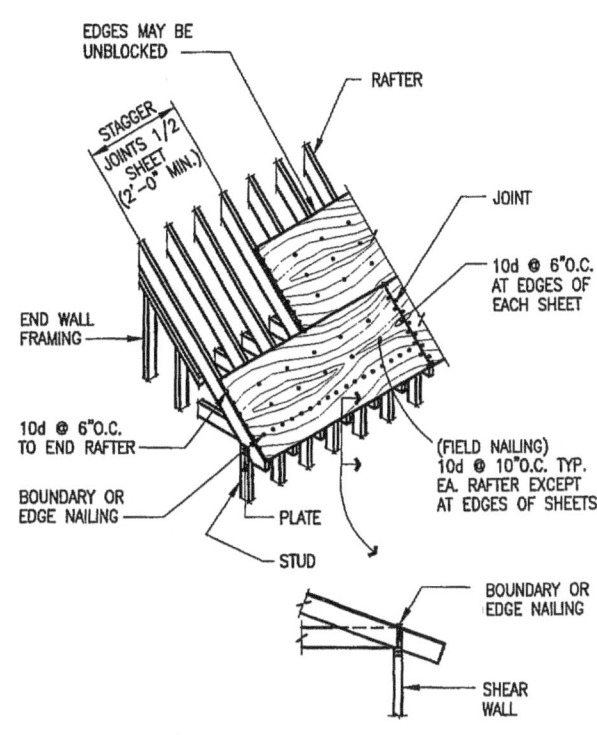

ROOF DIAPHRAGM SHEATHING
Figure 81

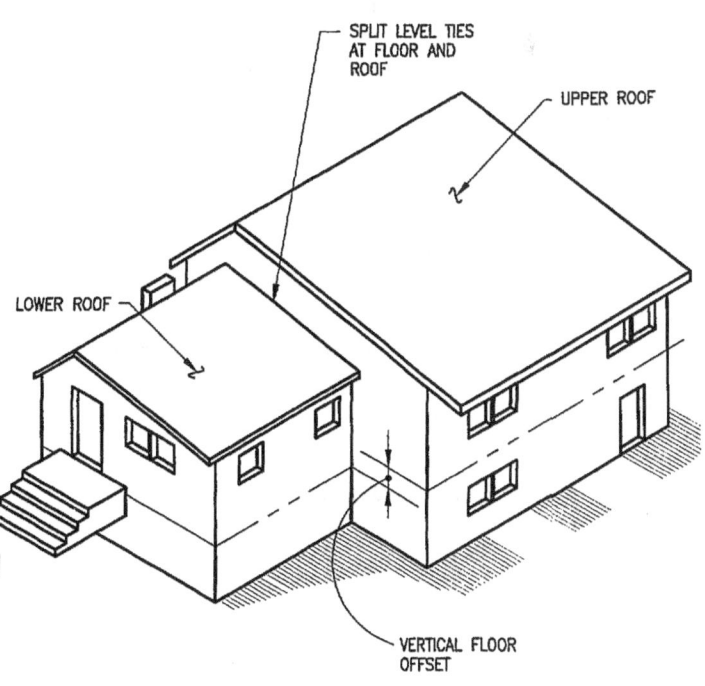

Figure 82

Other roof shapes are acceptable if precautions are observed to retain the characteristics of single plane surfaces and symmetry and to maintain continuity at breaks in the plane of the diaphragm. This includes such roof shapes or styles as pitched roofs, shed roofs, gables with both moderate and steep slopes, hips and valleys, and mansards as long as roof surfaces in the same plane remain unbroken and vertical offsets and discontinuities at wall boundary lines are avoided. See Table No. 4, page 48.

To function properly, diaphragms must conform to specific shape limitations that are controlled by the diaphragm ratios. See Figures 83 and 84. If the exterior wall locations are such as to make the span-width ratio unacceptable, interior shear walls must be used in conjunction with the outside walls so that acceptable span-width ratios are maintained. Each edge of the diaphragm should be supported by a shear wall. However, it is permissible to use shear walls which are not under the edge of a diaphragm by using collectors or drag struts connected to the shear walls. The most commonly used collectors are wall top plates or rim or header joists secured to the top plates. If rim joists act as collectors, they must be spliced for continuity.

Openings in roofs should be limited in number and moderate in size. In general, they should be restricted to one principal opening for a fireplace chimney and several smaller penetrations for plumbing vents, conduits and pipes. Large openings, as for skylights and atria, should have special nailing of the diaphragm sheathing to framing placed around the opening. Openings should be avoided at the corners of diaphragms.

At ridges or changes in slope it is necessary to tie the sections of the diaphragm together by nailing to blocking common to both planes and overlapping and nailing the framing members perpendicular to the ridge at the change in slope. See Figure 11.

Several common sheathing materials are available for roof diaphragms. Structural grade plywood or oriented strand board is recommended in high seismic risk areas. Where used, sheets should be oriented with the face grain perpendicular to the joists or rafters. See Figure 81.

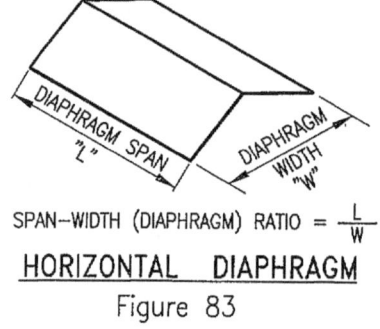

SPAN-WIDTH (DIAPHRAGM) RATIO = $\frac{L}{W}$

HORIZONTAL DIAPHRAGM
Figure 83

MAXIMUM DIAPHRAGM RATIO
(See also Table No. 3, Page 39)

Material	SEISMIC RISK	
	High	Moderate & Low
	Max. Span-Width Ratio	
1. Plywood * ** sheathing with edges blocked and unblocked	4 : 1	4 : 1
2. Diagonal sheathing	2 : 1	3 : 1
3. Straight sheathing	Not Recommended	2 : 1 Low Seismic R Area Only

* Wood Structural Panel
** Oriented Strand Board can be substituted for plywood

Figure 84

Blocking can be omitted along the long edges of the sheets; but, if it is omitted, metal ply clips along the unblocked edges should be used to prevent differential movement of the edges which could tear the roofing. To limit this type of damage or to create a stiffer diaphragm, blocking under all edges of Wood Structural Panel sheets with edge nailing of the sheathing to the blocking may be used. The short edges of abutting sheets must be supported by the same framing member with edge nailing along the common member for each sheet.

Diagonal sheathing, no longer commonly used, is an option for earthquake resistance for any seismic risk area and can be used as a substitute for plywood provided the nailing is as required by code. See Section 7 FLOORS for a discussion of this material. Straight sheathing should only be used as a diaphragm in low seismic risk areas.

In exposed beam ceilings, framing 1x or 2x tongue and groove straight sheathing over the tops of the beams may be used. A Wood Structural Panel overlay should be placed over the sheathing in high and moderate seismic risk areas. Edge nailing should be provided at the boundary of each Wood Structural Panel sheet. Field nailing should be located to avoid penetrating straight sheathing boards.

Skip-spaced boards used for attaching shingles form a very weak diaphragm and are not recommended. Spaced boards should only be used in low seismic risk areas that also have a low wind exposure. Skip spaced boards may not be used where prohibited by Code.

In order to form complete diaphragms, top plates of walls supporting roof diaphragms usually act as chords and/or collectors and should be continuous. The plates must be internailed and receive additional nailing in the lap length at splices. See Figure 85.

Top plates must provide continuity along the full widths and lengths of diaphragms. Roof sheathing must be nailed to rim joists parallel to framing or blocking inserted at the ends of the joists perpendicular to the walls, see Figure 86. Rim joists or blocking must be attached to the top plates. Prefabricated sheet metal framing clips can be used to secure rim joists or blocking to the plates.

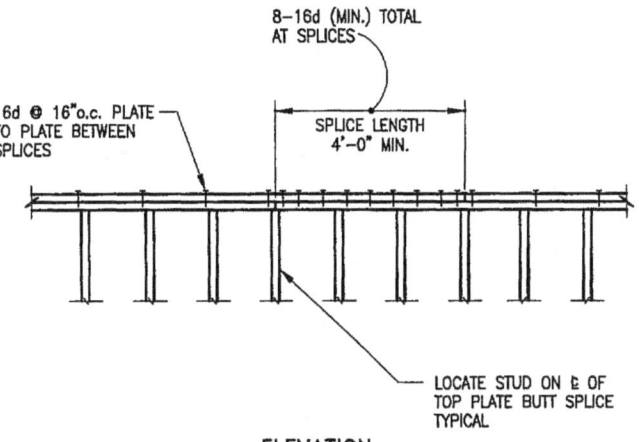

ELEVATION

TYPICAL TOP PLATE SPLICE

Figure 85

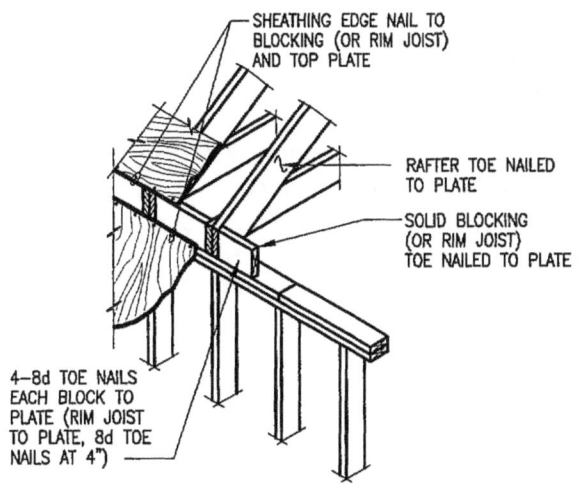

ROOF DIAPHRAGM CONNECTION TO SHEAR WALL

Figure 86

Earthquake forces in roof diaphragms can be delivered to shear walls that are outside the diaphragm limits by collectors that are connected to diaphragms. Collectors should run in straight continuous lines and be attached to the tops of shear walls. See Figure 87.

Collectors can be used to transfer earthquake loads developed in areas of roofs that project out from main roof diaphragms such as roofs or canopy projections over an open porch. Collectors or ties must be nailed to the projecting roofs and extend into and be nailed to the sheathing within the limits of the main roof diaphragms. See Figure 88.

In high seismic risk areas, minimum nailing of roof sheathing should be in accordance with code requirements. Roof sheathing should be fastened with nails with a diameter equivalent to common or box nails. When using a nailing gun to fasten plywood or oriented strand board, care must be exercised to avoid overdriving the nails. The gun operator should also confirm that the nails penetrate the framing members below. A new nail should be driven for each nail not properly installed. Adhesive applied to rafters under roof sheathing is not conventionally used. However, as for floors, adhesives will stiffen roof diaphragms and improve their ability to withstand earthquake shaking.

Where architectural expression requires roof configurations not specifically covered in the Guide, the builder should consult a design professional.

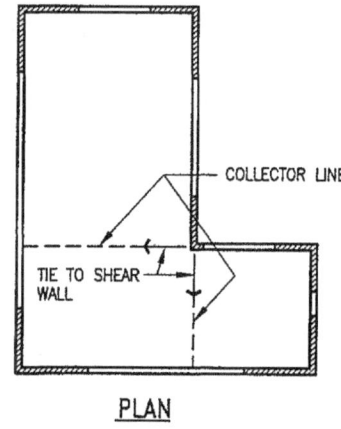

PLAN

Figure 87

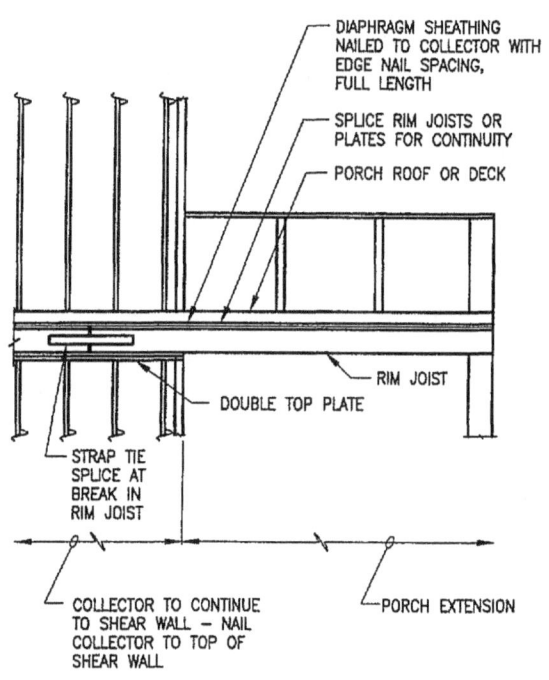

ELEVATION

TYPICAL COLLECTOR TIE AT PORCH EXTENSION

Figure 88

Some roof forms are not considered to be covered by the details included in this Guide. This does not imply that such forms cannot be made to be earthquake resistant. However, application of standard details shown in this Guide may not be appropriate and professional guidance is required. The following descriptions and illustrations show two forms not covered by the Guide.

1) Monitor or pop-up elevated portions of a roof that interrupt an otherwise complete diaphragm. See Figure 89. Monitor roofs are often supported on mullions with an all glass exterior and with no solid shear wall panels between roof levels.

2) Shed dormers that extend over nearly the full length or width of a roof plane especially on the sloping faces of gable roofs or hip framing. See Figure 90.

Other roof configurations excluded from the Guide are roofs with the following characteristics:

1) Large square or rectangular openings in roof planes for skylights or light shafts especially at corners.

2) Large openings in roofs over central atria or interior patios where the courtyard walls have all glass exposure, or where there are only column supports for the framing above.

3) Discontinuities at ridges where the roof sheathing does not extend to the break in the roof planes.

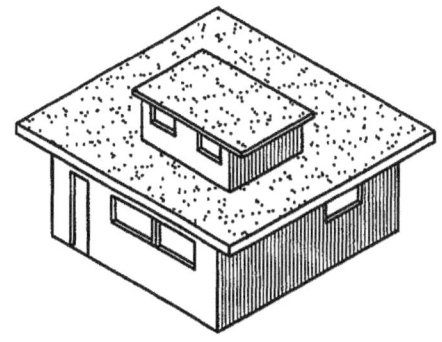

MONITOR ROOF

Figure 89

SHED ROOF

Figure 90

Table No. 4 Roof Shapes	
Flat Roof	
Mansard	
Shed Roof – Opposing Pitch	
Gable Roof	
Hip and Valley Roof	

10. MASONRY CHIMNEYS

Observations from the Northridge Earthquake as well as other earthquakes indicate that masonry chimneys, for reasons noted below, are not appropriate when used with light frame construction in high seismic risk areas unless properly reinforced, anchored and constructed. See Figure 91.

Masonry chimneys are usually heavy, rigid, brittle and may be highly susceptible to damage from earthquakes. Because of their weight and rigidity, they can become falling hazards and require special care when used with buildings such as wood framed structures in high and moderate seismic risk areas. See Figure 95a. When built monolithically with masonry buildings, chimneys will generally perform satisfactorily.

In high seismic risk areas, wood frame or steel stud framing for flue or chimney enclosures is recommended in lieu of masonry. See Figure 95b. Heavy finishes such as stone or masonry veneer should be avoided for chimneys where possible.

Where used, the dead load of the masonry should be supported independently on a substantial reinforced concrete spread footing or pad. A masonry chimney flue enclosure must have at least a minimum amount of horizontal and vertical reinforcing. The vertical reinforcing should run the full height without splices and each bar should be doweled to the footing. To prevent separation of the chimney from the building, anchor ties secured to the framing must be installed. These ties must be provided at every diaphragm level. See Figure 92. The ties must engage the diaphragm sheathing, be embedded in the masonry enclosure, and extend to and be bent around the vertical reinforcing on the far side of the chimney.

Minimum vertical reinforcing should consist of at least one vertical bar in each corner for chimneys of usual size. Chimneys that are large in plan may require more vertical reinforcing. Horizontal ties (at least #4 @ 18") should be placed around the vertical reinforcing. Sufficient grout space around the flue must be provided to fully embed the reinforcing. See Figures 92 and 93.

Vertical bars and ties must continue to the tops of the projection of chimneys above roofs. Grouting must also continue to the top and be continuous without interruption or voids.

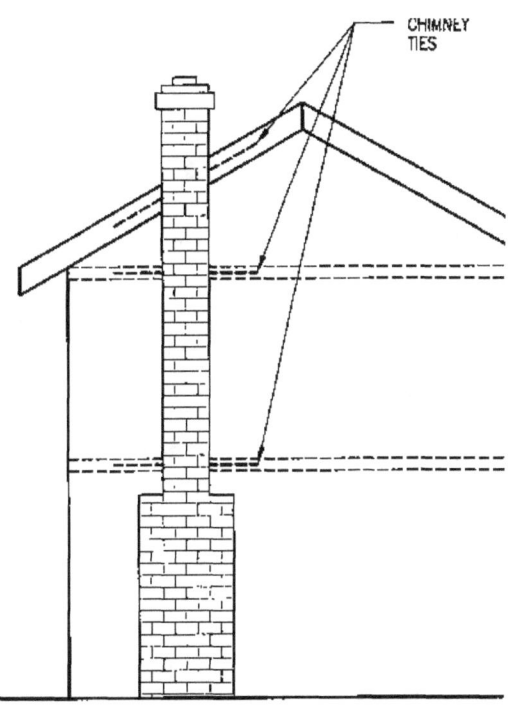

Figure 91

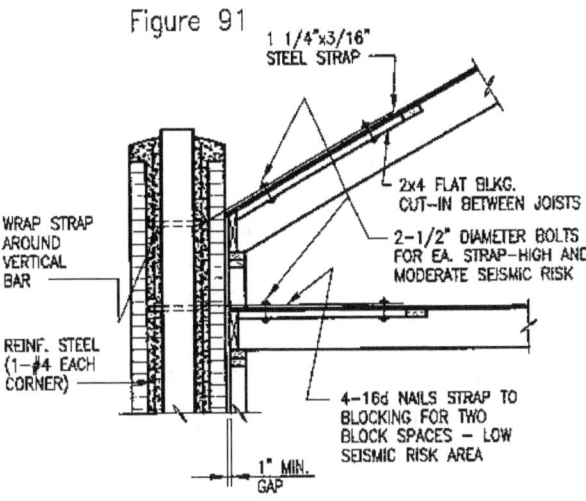

Figure 92

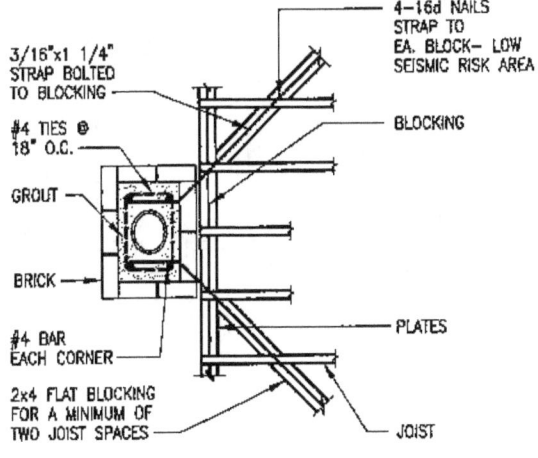

Figure 93

Strap anchors holding the chimneys transfer earthquake loads into the diaphragms. See Figures 92 and 93. Therefore, the straps should extend diagonally into the diaphragms and be bolted to the framing with 2-1/2" diameter bolts in high and moderate seismic risk areas. One strap should be located on each side of the chimney. Intermediate straps may be needed for large chimneys. The diaphragm sheathing should be nailed with 8-10d nails to the same member to which the strap is bolted. For low seismic risk areas, the straps may be nailed to the framing with 4-10d nails.

In the case of floor ties where the joists run at right angles to the straps, the straps should extend a minimum length into the diaphragm as follows:

High Seismic Risk — 4 joist spaces
Moderate Seismic Risk — 4 joist spaces
Low Seismic Risk — 2 joist spaces

When floor joists run parallel to the straps, the straps can be installed in the same manner as above or the straps may project from the face of the wall into the diaphragm and be bolted to the face of a joist or rafter.

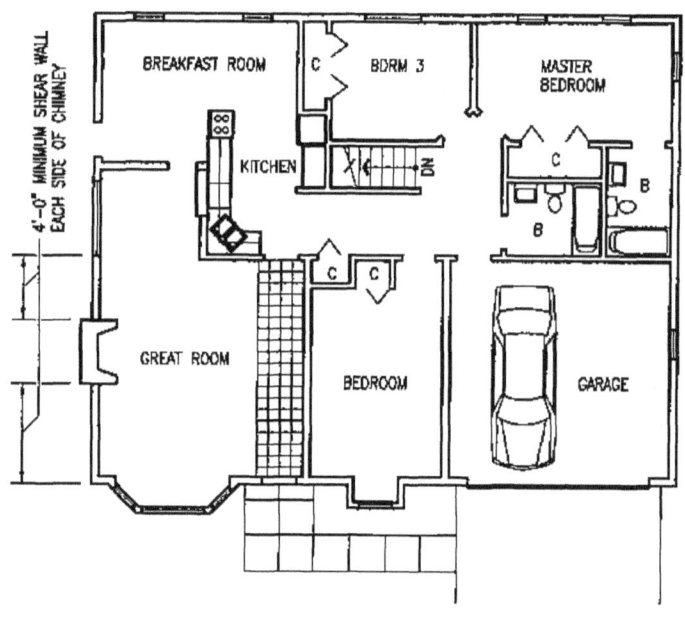

Figure 94

As noted before, anchor straps should be attach to each block as follows:

High seismic risk — 2-1/2" ⌀ bolts
Moderate seismic risk — 2-1/2" ⌀ bolts
Low seismic risk — 4-10d nails

Anchor straps prescribed by building codes have, at times, proven to be inadequate in earthquakes. For this reason it is important in high seismic risk areas to use only bolted straps or, better yet, to avoid masonry chimneys in favor of lighter weight chimney framing.

Because of the concentrated loads introduced into floor or roof diaphragms by earthquake loading from chimneys it is necessary to avoid openings in diaphragms near or adjacent to chimneys except as required for the flue or chimney structures. Openings should not occur adjacent to the anchor straps. Chimney openings in roof diaphragms should receive perimeter framing with edge nailing of the surrounding sheathing.

A minimum length of 4'-0" of shear wall should be located on each side of and immediately adjacent to a masonry chimney. See Figure 94. These walls should be solid sheathed without openings. Heavy veneers should not be attached to these shear walls. Chimneys that are reinforced and anchored in low seismic risk areas may not need adjacent shear walls, but better protection can be achieved if wall bracing is available in the walls adjacent to chimneys and windows or doors are avoided in the vicinity of chimneys.

Chimney loads will likely increase the forces in the adjacent shear walls. In narrow end walls, penetrations for windows or doors should be avoided or be kept to a minimum to provide as much solidly sheathed shear wall as is architecturally acceptable.

The UBC requires that every masonry chimney in high and moderate seismic risk areas be reinforced and anchored at each floor or ceiling line more than 6 feet above grade.

Figure 95a

Figure 95b

11. CONCRETE MASONRY

Walls of hollow concrete masonry are frequently used to resist earthquake forces in residential construction. In most respects the same principles apply as with wood and steel stud framed buildings. The masonry walls will act as shear walls. A complete earthquake resisting system includes roof and floor diaphragms, chords, collectors, shear walls and foundations. Because masonry walls are heavier than stud framing, it follows that inertial forces will be greater and the earthquake loads that must be resisted by the system will be larger. Because the forces to be resisted are larger, a greater demand is placed on the components of the resisting system. Therefore, care should be exercised in limiting openings in roof diaphragms and rigidly adhering to prescribed diaphragm ratios and diaphragm nailing requirements. Earthquake loads perpendicular to walls are more significant than for light frame construction. Therefore, the walls must be well anchored to the diaphragm at the top of the wall and to the foundation. See Figures 96 and 97.

Masonry dwellings covered by this Guide are limited to one and two story construction. Figure 98 shows a one story building. Masonry walls can be full height from ground floor to roof; or wood or steel framing may continue above the masonry from the floor line to the roof for two story construction. Except for veneer, gravity loads from masonry should never be supported by wood framing. It is desirable to keep floor plans regular and rectangular. It is also good practice to locate shear walls so as to keep the ratio of length to width of horizontal diaphragms (diaphragm ratio) below those shown in Table 3, page 39.

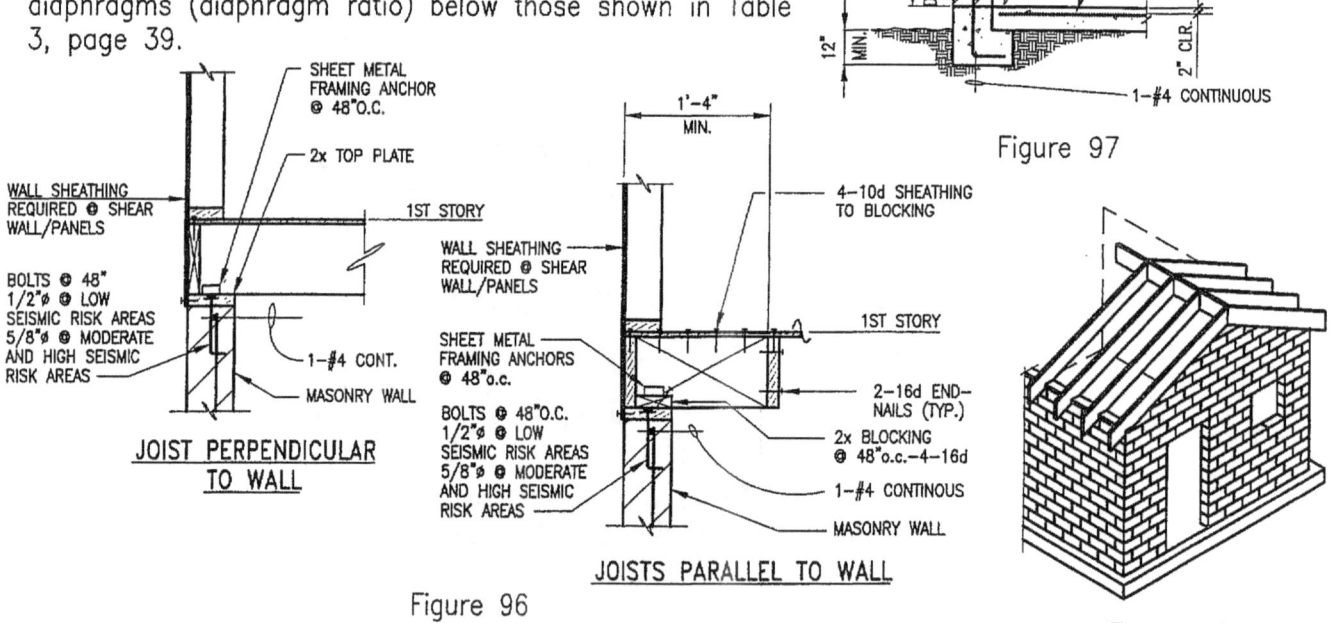

Figure 96

Figure 97

Figure 98

A 4'-0" width of shear wall should be maintained in each of the exterior elevations of the building. For unreinforced shear walls, where permitted, the minimum width should be 6'-0". To reduce the possible detrimental effects of rotation of the diaphragm, shear walls should be located near corners of the building in mutually perpendicular directions. Maximum distance between shear panels is 35 feet. See Figure 99. If the walls intersect at the corners, at least one vertical reinforcing bar should be placed at the corner and horizontal reinforcing bars, where needed, should extend into the corner and be lap spliced around the corner, See Figure 105.

Foundations for concrete masonry walls should be continuous wall footings or slabs-on-grade with thickened edges. See Figure 100. Footings should be continuous around the exterior of the building and under interior shear walls. Footings should be reinforced in high seismic risk areas, and reinforced masonry walls should be doweled to the footings. See Figure 101. Unless required by the building code or local regulations, footings need not be reinforced in low seismic risk areas.

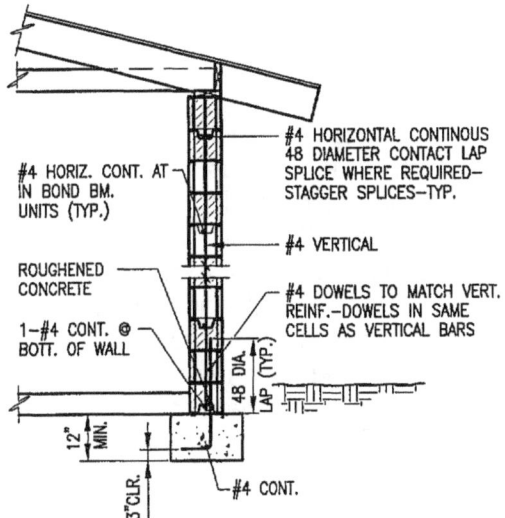

Figure 100

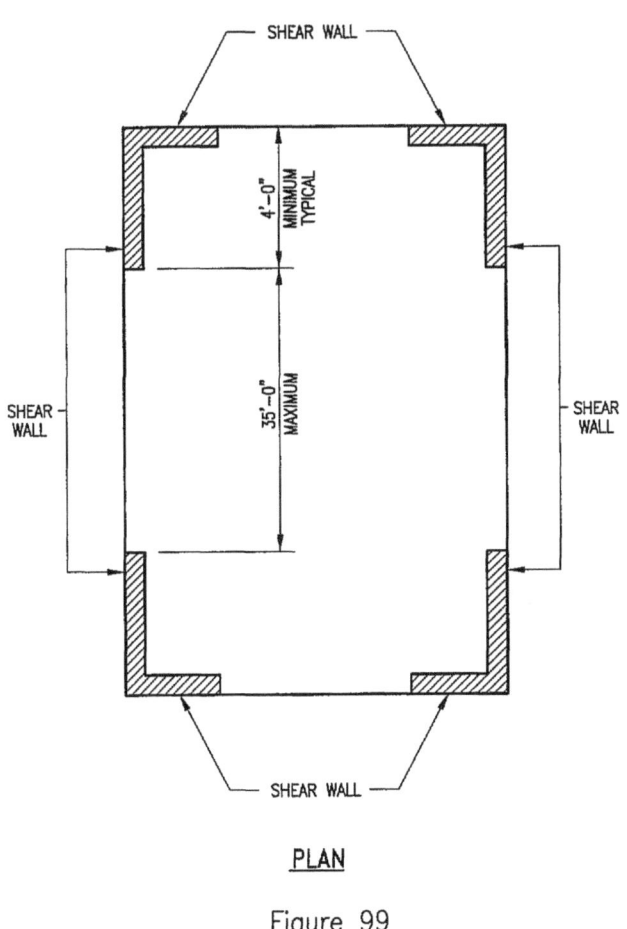

PLAN

Figure 99

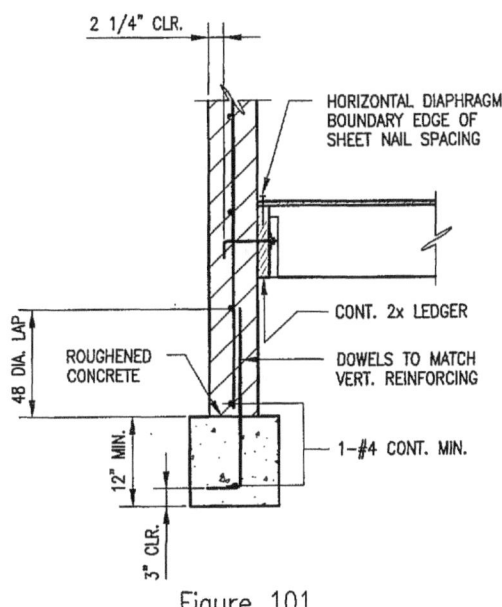

Figure 101

(53)

Builders have a choice of bond patterns for masonry walls. The pattern most often used is common or running bond. See Figure 102a. In common bond, units in adjacent courses overlap. Where the overlap is at least 25% of the length of the units, the wall is considered to be running bond.

Running bond is the most effective pattern for resistance to earthquakes and is especially desirable for high seismic risk areas. For this pattern, the vertical joints are offset in each course thus bonding the masonry units together for the full height of the wall and increasing its strength.

Where there is no overlap of units, the pattern is referred to as stack bond. See Figure 102b. In stack bond masonry, the vertical joints line up for the full height of the wall which results in a weak vertical plane within the wall. For high and moderate seismic risk areas, reinforcing of stack bond masonry is desirable and open end units with walls grouted solid are recommended in order to compensate for the inherent weakness of the stack bond pattern. Even in low seismic risk areas a minimum amount of horizontal reinforcing across the vertical joints is desirable. Horizontal reinforcing may be supplied by joint reinforcing @ 16" o.c. placed in the mortar bed joints.

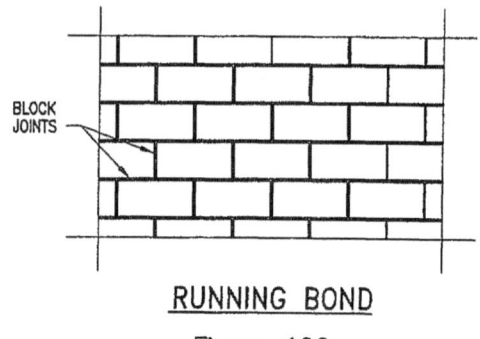

RUNNING BOND

Figure 102a

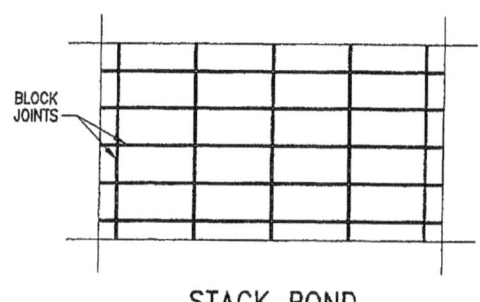

STACK BOND

Figure 102b

In high seismic risk areas, masonry walls require vertical and horizontal reinforcing. See Figure 103. Vertical bars must extend to the the top of the masonry wall including the parapet, if there is one. Where single length vertical bars cannot be used, lap splices of at least 48 bar diameters are required. Vertical bars should be in the same cell as the footing dowels and should be wire-tied to horizontal bars at the top and bottom of the wall and at intermediate locations. However, where foundation dowels are misplaced, vertical bars may be offset up to a maximum of 8". Vertical reinforcing should be located at every wall intersection and door and/or window jamb. Metal positioners can be used to hold the vertical bars in place. In moderate seismic risk areas, vertical bars need only be placed at corners of intersecting walls, at each end of shear walls and at door and window jambs. The vertical reinforcing bars should be doweled to footings.

In high and moderate seismic risk areas, horizontal reinforcing bars should be placed at the diaphragm levels around the perimeter of the floor and roof diaphragms to resist chord tension forces. The chord bars should be as long as possible to minimize splices and be spliced at corners.

Except for chord bars, horizontal bars can be replaced by ladder wire reinforcing placed in horizontal mortar joints usually at 16" o.c.. Required horizontal reinforcing bars should be placed in bond beam units and not in the mortar joints. See Figures 104 and 105. Reinforcing requirements should be verified with governing building codes and/or local practices.

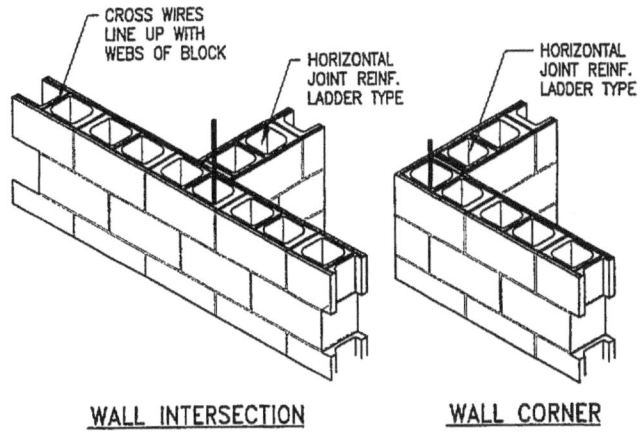

Figure 104

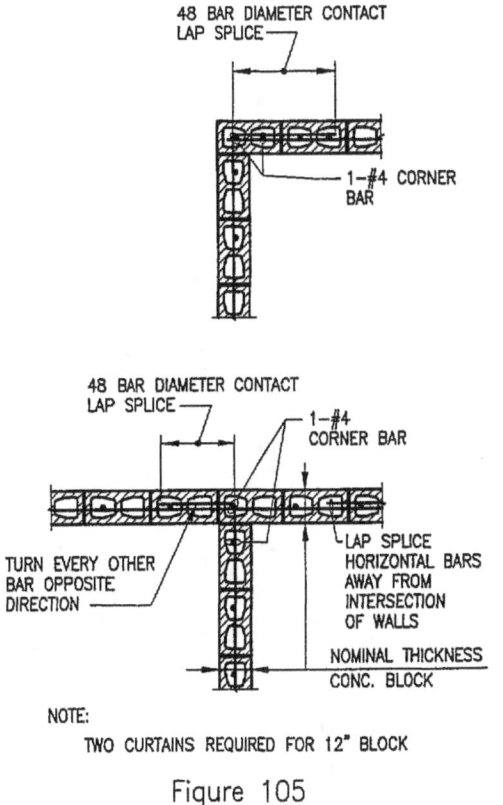

Figure 105

Figure 103

Every cell with vertical reinforcing must be fully grouted. See Figure 106. Horizontal bars in bond beams must also be grouted. In high seismic risk areas, consideration should be given to grouting all the cells.

As described before, and now described in more detail, where masonry walls act as shear walls, it is necessary that the floor and roof diaphragms be connected to the walls. Where the joists or rafters rest on top of the wall, the diaphragm sheathing must be nailed to blocking or joists that are nailed to the wall top plates. The top plates must be secured to the top of the wall with bolts that are grouted into the cells of a bond beam. See Figures 107 and 108 for conditions at roof framing.

If diaphragm sheathing abuts the face of a wall it must be nailed to a ledger. Ledgers must be bolted to the face of a wall with bolts placed in grouted cells. See Figures 109 and 110. In high and moderate seismic risk areas, wall ledger bolts should be a minimum of 5/8" diameter and should be headed or have hooks at the embedded non-threaded end. Spacing should not exceed 48" on center. For low seismic risk areas, ledger bolts used to resist vertical loads will generally be adequate to resist earthquake loads as well. In any case, ledger bolts must be spaced as required to support dead and live loads.

Ledgers and top plates should have a bolt not less than 9" from the ends of each piece including butt splices. Plates and ledgers should be installed with pieces as long as possible to minimize splices.

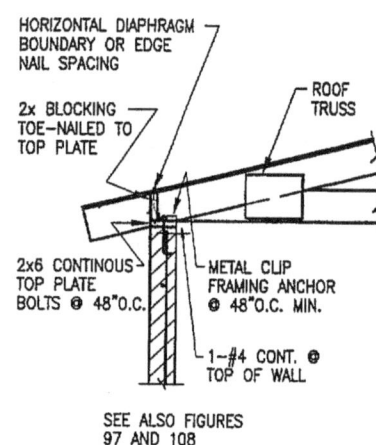

Figure 107

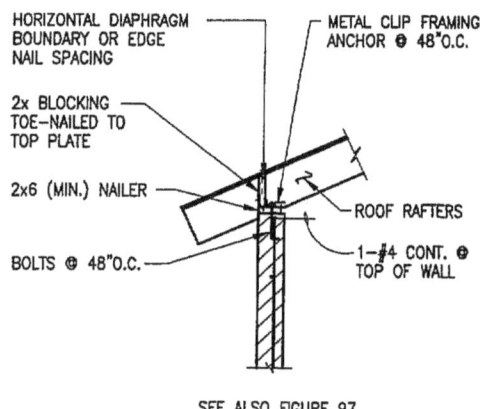

Figure 108

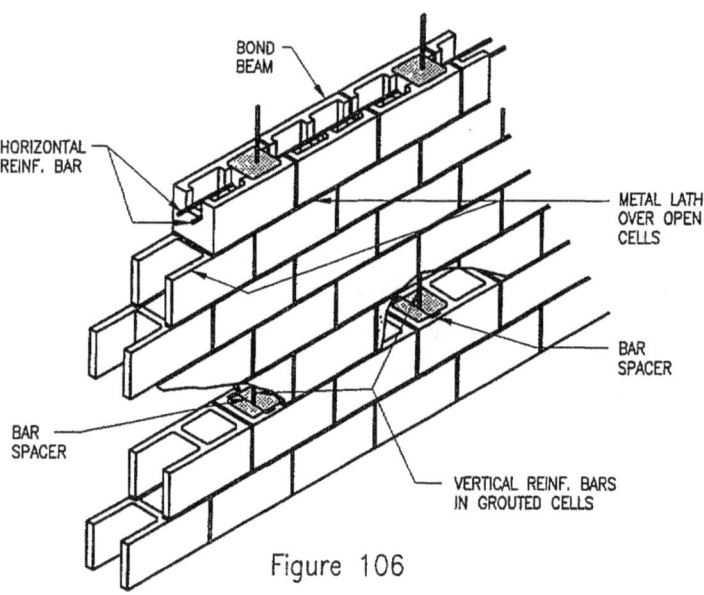

Figure 106

In high seismic risk areas where diaphragm sheathing nailed to wall ledgers is the only sheathing-to-wall connection, it has proven to be inadequate for providing lateral support for the walls. Nails may pull through the sheathing and/or ledgers may fail in bending across the grain of the wood (crossgrain bending), and allow the wall to fall away.

In high and moderate seismic risk areas, prefabricated anchor straps should be used to anchor walls to horizontal diaphragms so as to transfer seismic forces without allowing crossgrain bending in ledgers. Anchors should be spaced a maximum of 4'-0" on center completely around the exterior of the building. Interior masonry walls should also be anchored to floor and roof framing in the same manner. The anchor straps must be embedded in grouted cells and be attached to joists or blocking. See Figures 109 and 110. They should be carefully located and in line with and parallel to joists, rafters or blocking. The straps should be attached by nails to the top or bottom edges of framing members or by bolts or nails to the face of framing members. Where floor and roof framing run parallel with the face of the walls, anchor straps may be nailed to the underside of solid blocking cut in between joists or rafters. See Figure 111. Straps should be long enough to extend from the face of the wall a minimum of two joist or rafter spaces. Twisted or riveted strap anchors should be avoided. Floor and roof sheathing should receive extra nailing to the members holding anchor strap.

Local regulations should be consulted for bracing of masonry walls for wind loads during construction.

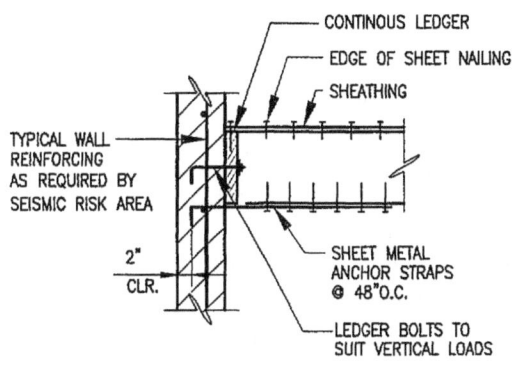

JOIST PARALLEL TO WALL

Figure 109

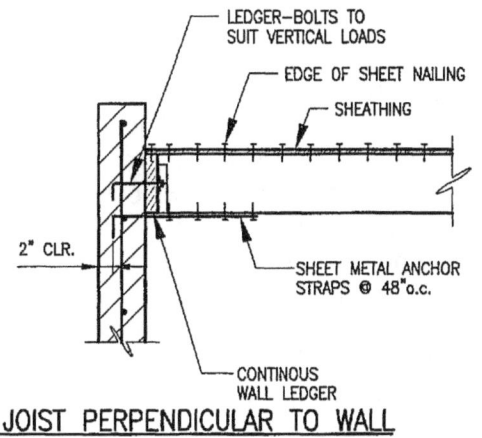

JOIST PERPENDICULAR TO WALL

Figure 110

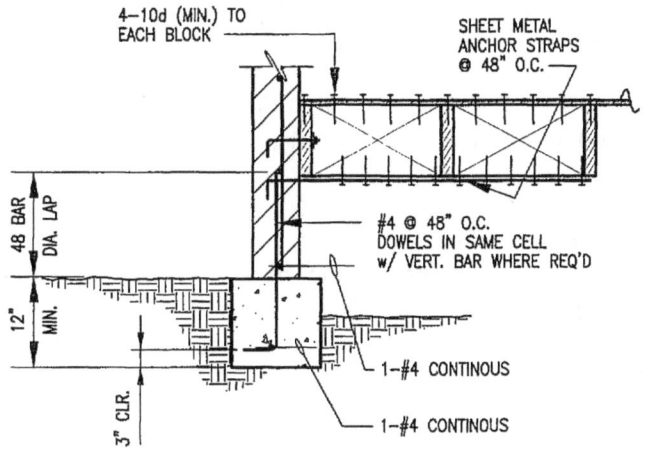

Figure 111

12. CLAY MASONRY

As with concrete masonry construction, walls built with clay brick masonry units may be used in residential construction to act as shear walls to resist earthquakes. Location, spacing, width, height to width ratios for shear walls and foundation requirements are the same as for concrete masonry. Anchorage of walls for loads perpendicular to the face of walls and wall ledger details are also the same. Detailing and reinforcing requirements for hollow units are the same as for hollow concrete masonry units. Refer to Section 11 CONCRETE MASONRY.

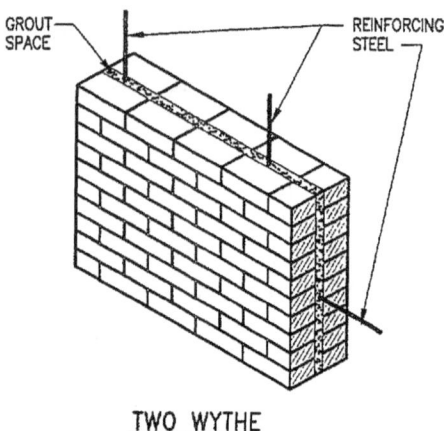

TWO WYTHE GROUTED REINFORCED

Figure 112

Brick shear walls built with solid units must have reinforcing in high risk areas. Reinforced walls are generally constructed with two wythes of brick separated by a grout space. See Figures 112 and 114. The vertical and horizontal reinforcing is positioned within the grout space and fully embedded in grout although, as with hollow masonry, the grout space need not be solidly filled. Additional vertical wall reinforcing should be placed at the jambs of door and window openings and a horizontal bar should be located at door and window heads. The two wythes of brick should be held together with metal wall ties or ladder wire joint reinforcing placed in the mortar joints. Ties should be provided for each four square feet of wall surface. The maximum vertical spacing of wall ties or wire joint reinforcing should not be more than 24"

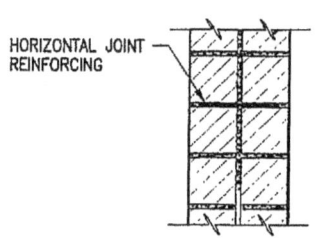

TWO WYTHE NOT REINFORCED

Figure 113

Empirical design rules for solid masonry construction can be found in ACI 530/ASCE 5/TMS402 Building Code Requirements for Masonry Structures and in model codes. According to these rules, shear walls must have a minimum thickness of 8". Shear walls must conform to a height to thickness ratio of 18 or less. Reinforcing is as required by empirical design rules. Empirical design rules are not generally applicable in high seismic risk areas.

Brick shear walls in moderate and low seismic risk areas can be two wythe cavity wall construction and need not be reinforced unless required by the governing code. See Figures 113 and 114. The brick courses should be laid up in running bond with metal ties placed between the wythes as for grouted walls.

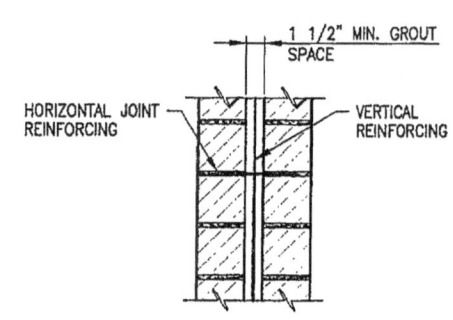

TWO WYTHE GROUTED REINFORCED

Figure 114

13. MASONRY AND STONE VENEER

Masonry and stone is often used as veneer attached to wood stud or steel stud backing. Satisfactory performance of veneer can be achieved if the support system provides an adequate foundation, strong and stiff backing, proper attachment of the veneer to the backing and good workmanship. Mortar is an important ingredient of veneer and all mortar joints should be completely filled. Mortar joints should be tooled to increase the water tightness of the wall. A portland cement-lime mortar is recommended.

Masonry or stone veneer should be supported on a concrete foundation such as the wall footing, an independent foundation or noncombustible supports such as steel lintel angles. See Figure 115.

For the best performance in earthquakes, it is recommended that a Structural Wood Panel backing be used on the veneer side of the supporting walls and gypsum board sheathing be used on the room side. See Figures 115 and 116. Structural Wood Panels should be exterior grade and not less than 3/8" in thickness. For low seismic risk areas exterior grade gypsum board sheathing 1/2" thick may be used for sheathing on the veneer side of the walls. Non-corroding nails or screws should be used to fasten the veneer anchors to the sheathing and studs. Steel studs should be 18 gage minimum with G90 zinc galvanized coating.

Veneer should have a 1" space between the veneer and the wall sheathing. See Figure 115. This space may be filled with grout for better performance. Building paper should be attached directly to the face of the sheathing for wood studs or self-furring insulation board can be attached directly to the face of steel studs.

Veneer must be anchored with sheet metal or wire ties fastened with screws penetrating through the building paper and sheathing and into the studs. Ties should project into mortar joints. In high seismic risk areas, veneer should have joint reinforcing engaging the ties in the mortar joints of the veneer. Ties should be galvanized.

A small diameter wire placed in the horizontal mortar joint is recommended for moderate and high seismic risk areas. The wire should be placed in the same mortar joint with the veneer anchors. This detail is recommended for stack bond in all seismic risk areas.

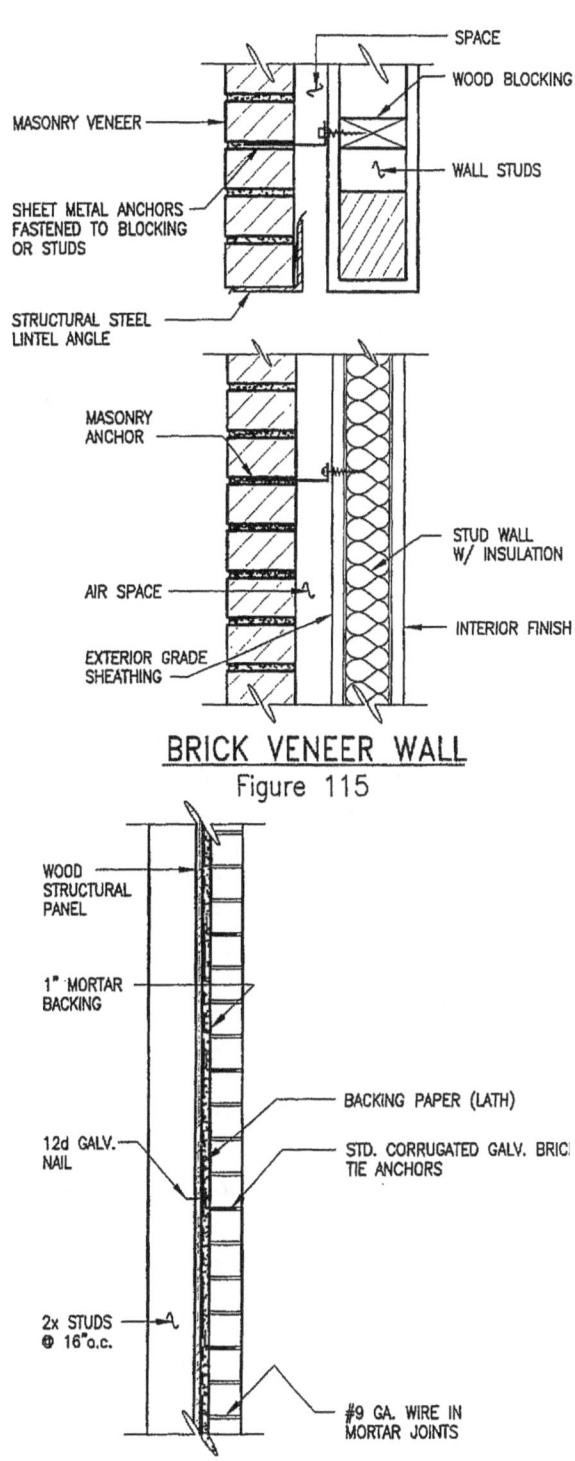

BRICK VENEER WALL
Figure 115

Figure 116

14. BUILDING CODES AND REFERENCES

Building code provisions are developed, adopted and enforced by jurisdictions to safeguard the health, safety, and welfare of building occupants. Building code requirements that relate to earthquakes are intended to prevent collapse and local failures that might endanger the safety of occupants and, to a lesser extent, mitigate economic loss due to property damage.

Building codes specify procedures that should be followed in the construction of buildings and their critical structural components. Specific attention is given to shear walls or other types of bracing, connections of floors and roofs to walls, connections of walls to foundations, and anchoring of major appliances such as water heaters. Codes require that all structural components that resist seismic forces be connected with adequate strength and stiffness to provide a continuous load path to the foundation.

Seismic requirements in codes are developed from information provided by research and lessons learned from past earthquakes. Risk and severity of earthquakes vary throughout the United States. The country is divided into seismic risk areas based on estimated frequency and magnitude of earthquake activity. Because of the expected severity of earthquakes in high seismic risk areas, provisions for earthquake resistance in these areas are more stringent than are required elsewhere.

As described in the INTRODUCTION, there are three model codes being used in the United States regulating the design and construction of buildings. They are the Uniform Building Code (UBC), the BOCA National Building Code (NBC) and the Standard Building Code (SBC). They are published by different organizations and are used in areas of the country as follows:

- The BOCA National Building Code (NBC) — northeastern states
- The Standard Building Code (SBCCI) — southern states.
- The Uniform Building Code (UBC) — states west of the Mississippi.

In addition to the three codes noted above, the Federal Emergency Management Agency (FEMA) has published the "NEHRP (National Earthquake Hazard Reduction Program) Recommended Provisions for the Seismic Regulations of New Buildings" which also contains provisions for residential seismic resistant construction. The NEHRP document serves as the basis for the seismic provisions of the NBC and SBC codes and will be used as a basis for the seismic aspects of the IBC (International Building Code).

To consolidate and standardize the requirements for detached one-and two-family dwellings, the CABO One and Two Family Dwelling Code was created by the three model code agencies. However, not all State and local governments have adopted the CABO One-and Two-Family Dwelling Code. In the following discussion of codes, emphasis will be placed on the seismic requirements of the CABO One and Two Family Dwelling Code. In high seismic risk areas, local jurisdictions using the UBC may not recognize the CABO One and Two Family Dwelling Code and require the UBC provisions to be observed.

The International Code Council (ICC) intends to publish a new building code in the year 2000. This code will be known as the "International Building Code" (IBC). The new code is intended to replace the three model codes. The ICC is also making recommendations to amend and revise the CABO One and Two Family Dwelling Code or replace the CABO Code with the International Residential Code (IRC). When revised and adopted, the new residential building code should be universally applicable throughout the United States. While the IRC document draws heavily from the CABO One and Two Family Dwelling Code, FEMA has determined that the CABO Code is not substantially equivalent to the NEHRP document in high seismic risk areas. In these areas, the appropriate model code should be used.

All model codes, including the CABO One and Two Family Dwelling Code, are revised periodically so that new regulations can be incorporated. State and local governments usually adopt the most current version of each code and, in some cases, with modifications. Home builders should ask their local building code jurisdiction or plan review agency about the specific codes which may apply to one-and two-family dwellings in their area.

Home builders should note that the codes set minimum requirements. Builders may always choose more conservative construction procedures to provide better protection against earthquakes. As noted before, in some instances, the recommendations included in the Guide may exceed the minimum code requirements because experience has demonstrated that superior performance in earthquakes can be achieved by simple, inexpensive details. In high seismic risk areas, the use of construction practices that exceed the minimum code levels may be used as a positive sales incentive.

Examples of CABO One and Two Family Dwelling Code requirements for basement walls and masonry construction by seismic risk areas are shown in Tables 6 & 7 on page 66. They include requirements for masonry and concrete foundation walls subjected to pressure equal to or less than exerted by backfill having an equivalent fluid weight of 30 pounds per cubic foot when located in seismic risk areas, or subjected to unstable ground conditions.

The CABO One and Two Family Dwelling Code also includes the following requirements:

1. All masonry or concrete chimneys that are in high seismic risk areas and that extend more than 7 feet above the last contact with the structure shall be reinforced. The code also requires that masonry or concrete chimneys in high seismic risk areas shall be anchored at each floor, ceiling, or roof line more than 6 feet above grade, except when constructed completely within the exterior walls.

2. Appliances fixed in position shall be fastened in place. Water heaters that have nonrigid water connections and are over 4 feet in height from the base to the top of the tank case shall be anchored or strapped to the building to resist horizontal displacement due to the earthquake motion. See Figure 118 on page 66.

Examples and comparisons of requirements for bracing and sheathing of wood framed walls, as included in the various codes referred to in the Guide, are shown in the following tables.

1. Table 8 on page 68 shows the minimum bracing (shear) wall panel types and their fastenings.

2. Tables 9 on page 69, 10 on page 70 and 11 on page 70 tabulate, for the various codes, where the bracing wall panel types shown in Table 8 may be used and other limiting requirements.

Reference Standards:

1. 1994 and 1997 Uniform Building Code (UBC). International Conference of Building Officials, 5360 South Workman Mill Road, Whittier, CA 90601-2298

2. 1994 and 1997 Standard Building Code (SBC). Southern Building Code Conference International, Inc., 900 Montclair Road, Birmingham, AL 35213-1206

3. 1993 and 1997 National Building Code (NBC), (BOCA). Building Officials and Code Administrators International, Inc., 4051 W. Flossmor Road, Country Club Hills, IL 60478-5795

4. 1995 CABO One and Two Family Dwelling Code. The Council of American Building Officials, 5203 Leesburg Pike, Falls Church, VA 22041

5. NEHRP Recommended Provisions for the Development of Seismic Regulations for New Buildings, 1994 and 1997 Editions. Building Seismic Safety Council, Washington, DC 20005

6. APA Residential Construction Guide. American Plywood Association, P.O. Box 11700, Tacoma, WA 98411-0700

7. 1997 supplement to the CABO One and Two Family Dwelling Code, the Council of American Building Officials, 5203 Leesburg PIKE, Falls Church, VA 22041

8. Prescriptive Method for Residential Cold-Formed Steel Framing – August 1997, prepared by the NAHB Research Center for HUD and AISI.

9. Low-Rise Construction Details and Guidlines – AISI, June 1993

10. Building Code Requirements for Masonry Structures (ACI 530-95/ASCE 5-95/TMS 402-95), American Society of Civil Engineers, 1801 Alexander Bell Dr., Reston, VA 20191-4400.

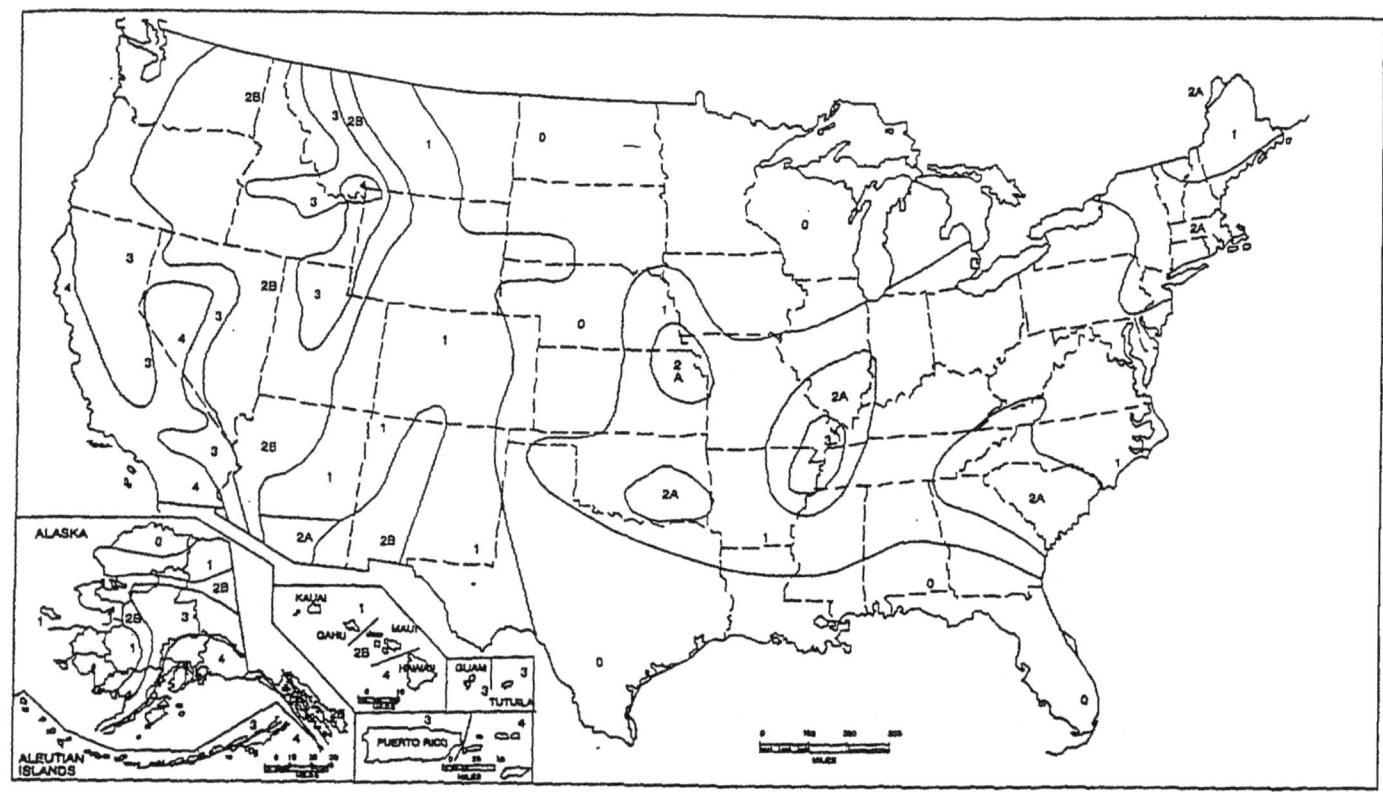

FIGURE 16-2—SEISMIC ZONE MAP OF THE UNITED STATES
For areas outside of the United States, see Appendix Chapter 16.

Figure 117

Reproduced from the 1997 Edition of the *Uniform Building Code.*™

Table No. 5
Seismic Risk Areas

SEISMIC RISK AREA	SBCCI BOCA	CABO UBC (ZONE)	MAPS PREPARED FOR 1997 NEHRP RECOMMENDED PROVISIONS FOR NEW NEW BUILDINGS	WHAT'S EXPECTED
High	D, E, F	2B, 3, 4	$S_{DS} \geq 0.5g$ $S_{D1} \geq 0.2g$	Ductile Behavior Significant seismic detailing recommended
Moderate	B, C	1, 2A	$S_{DS} \geq 0.167g < 0.5g$ $S_{D1} \geq 0.067g < 0.2g$	Full seismic resisting system with complete load path
Low	A	0	$S_{DS} < 0.167g$ $S_{D1} < 0.067g$	Meets wind load resisting requirements and min. component interconnection requirements

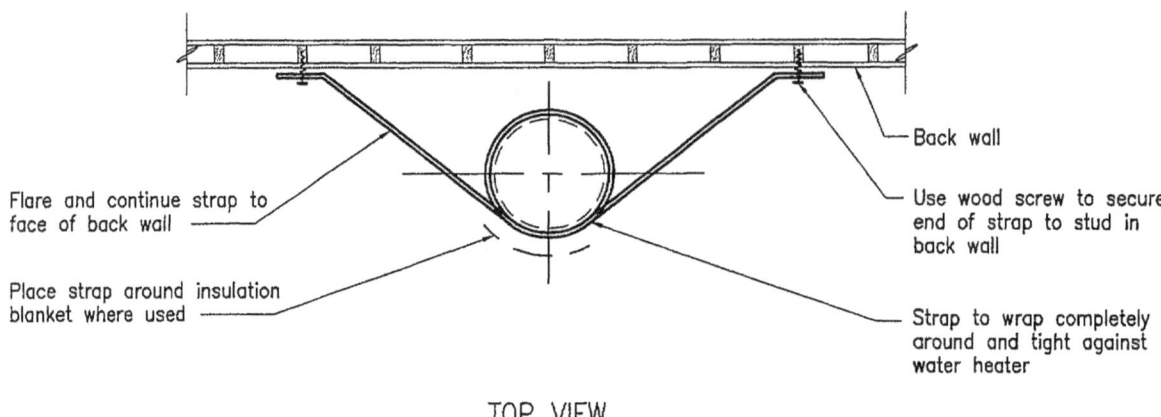

TOP VIEW

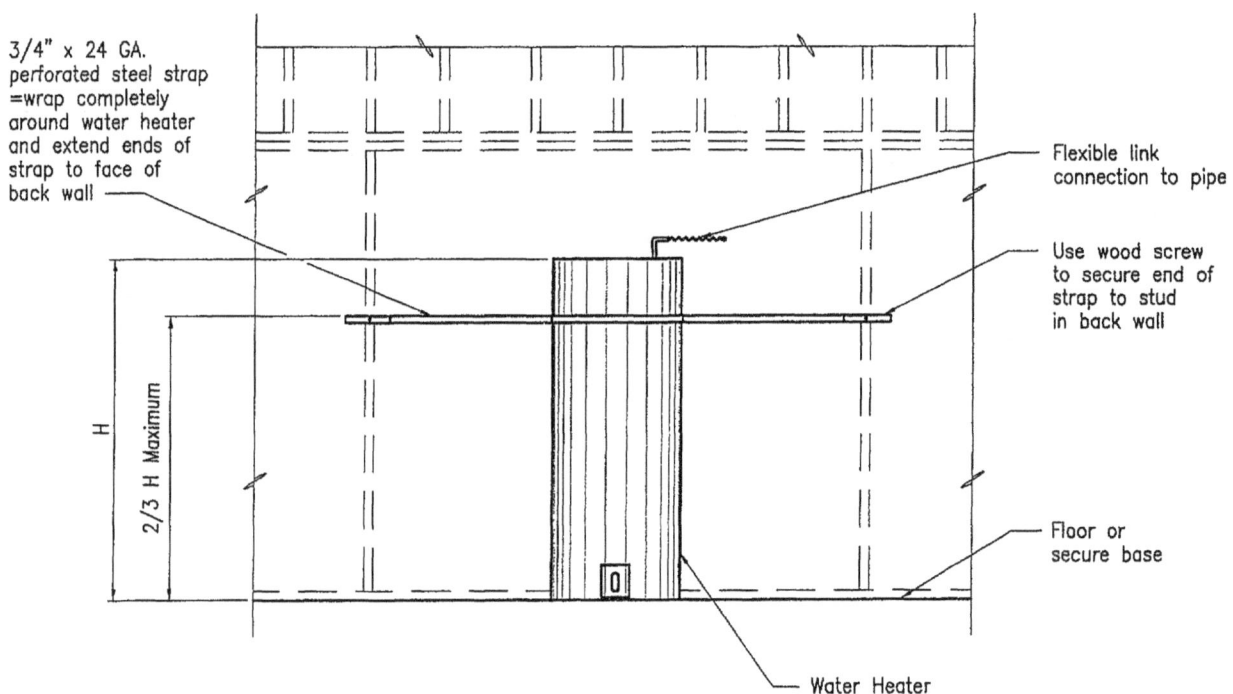

Note:
High Seismic Zones should have two straps – top and bottom

FRONT VIEW

WATER HEATER SECURED TO WALL

Figure 118

Table No. 6 - CABO ONE AND TWO FAMILY DWELLING CODE

MINIMUM THICKNESS ALLOWABLE DEPTH OF UNBALANCED FILL FOR UNREINFORCED MASONRY AND CONCRETE FOUNDATION WALLS [1,2] WHERE UNSTABLE SOIL OR GROUND WATER CONDITIONS DO NOT EXIST IN LOW SEISMIC RISK AREA

Foundation Wall Construction	Nominal Thickness [3] (inches)	Maximum Depth of Unbalanced Fill [1] (feet)
Masonry of Hollow Units, Ungrouted	8 10 12	4 5 6
Masonry of Solid Units	6 8 10 12	3 5 6 7
Masonry of Hollow or Solid Units, Fully Grouted	8 10 12	7 8 8
Plain Concrete	6 [4] 8 10 12	6 7 8 8
Rubble Stone Masonry	16	8
Masonry of hollow units reinforced vertically with No. 4 bars and grout at 24 inches on center. Bars located not less than 4 ½ inches from pressure side of wall.	8	7

For SI: 1 inch = 25.4mm, 1 foot = 304.8mm.

[1] Unbalanced fill is the difference in height of the exterior and interior finish ground levels. Where an interior concrete slab is provided, the unbalanced fill shall be measured from the exterior finish ground level to the top of the interior concrete slab.
[2] The height between lateral supports shall not exceed 8 feet.
[3] The actual thickness shall not be more than ½ inch less than the required nominal thickness specified in the table.
[4] Six-inch plain concrete walls shall be formed on both sides.

Table No. 7   CABO ONE AND TWO FAMILY DWELLING CODE

REQUIREMENTS FOR MASONRY OR CONCRETE FOUNDATION WALLS SUBJECTED TO NO MORE PRESSURE THAN WOULD BE EXERTED BY BACKFILL HAVING AN EQUIVALENT FLUID WEIGHT OF 30 POUNDS PER CUBIC FOOT LOCATED IN MODERATE TO HIGH SEISMIC RISK AREA SUBJECTED TO UNSTABLE SOIL CONDITIONS

Material Type	Height of Unbalanced Fill in Feet	Length of Wall Between Supporting Masonry or Concrete Walls in Feet	Minimum [2] Wall Thickness in Inches [3]	Required Reinforcing	
				Horizontal Bar in Upper 12 inches of Wall	Size and Spacing of Vertical Bars
Hollow Masonry	4 or less	unlimited	8	not required	not required
	more than 4	design required	design required	design required	design required
Concrete or Solid Masonry [4]	4 or less	unlimited	8	not required	not required
	more than 4	less than 8	8	2-No. 3	No. 3 @ 18" o.c.
	8 or less	8 to 10	8	2-No. 4	No. 3 @ 18" o.c.
	8 or less	10 to 12	8	2-No. 5	No. 3 @ 18" o.c.
	more than 8	design required	design required	design required	design required

For SI: 1 inch = 25.4mm, 1 foot = 304.8mm, 1 pound per cubic foot (pcf) = 0.1572 kN/m³.

[1] Backfilling shall not be commenced until after the wall is anchored to the floor.
[2] Thickness of concrete walls may be 6 inches, provided reinforcing is placed not less than 1 inch or more than 2 inches from the face of the wall not against the earth.
[3] The actual thickness shall not be more than ½ inch less than the required thickness specified in the table.
[4] Solid masonry shall include solid brick or concrete units and hollow masonry units with all cells grouted.

MINIMUM BRACING WALL PANEL TYPES

Wall Panel Type	Sheathing Type	Minimum Thickness	Maximum Stud Spacing	Fastener Type	Fastener Spacings	
					Panel Edge Nailing	Intermediate Bearing Points
1	Continuous diagonal braces let into top and bottom plates and intervening studs, placed at an angle not more than 60 degrees or less than 45 degrees from the horizontal	1 x 4 nominal	No limit	8d common nails	(2) at each stud and plate, face nail	(2) at each stud and plate, face nail
2	Diagonal wood boards	5/8" net x 6" nominal	24" o.c.	8d common nails	(3) per board	(2) per board
				8d box nails	(5) per board	(3) per board
		5/8" net x 8" nominal	24" o.c.	8d common nails	(4) per board	(3) per board
				8d box nails	(6) per board	(4) per board
3	Fiberboard	7/16"	16" o.c.	No. 11 gage, 1-½ inch long, 7/16 inch head, galvanized roofing nails	3" o.c.	6" o.c.
4	Wood structural panels	5/16"	16" o.c.	6d common nails	6" o.c.	12" o.c.
		3/8"	24" o.c.	6d common nails	6" o.c.	12" o.c.
5	2-M-W particle board sheathing	3/8"	16" o.c.	6d common nails	6" o.c.	12" o.c.
		7/16"	24" o.c.	6d common nails	6" o.c.	12" o.c.
	2-M-1, 2-M-2 or 2-M-3 particle board sheathing *	½"	16" o.c.	6d common nails	6" o.c.	12" o.c.
6	Gypsum sheathing	½"	16" o.c.	No. 11 gage, 1-3/4 inch long, 7/16 inch head, diamond head, galvanized nails	4" o.c.	4" o.c.
7	Gypsum wallboard or veneer - 4 foot wide sheets	½"	24" o.c.	1-3/8 inch drywall nail (ASTM C514)	8" o.c.	8" o.c.
8	Portland cement plaster on expanded metal or woven wire lath	7/8"	16" o.c.	No. 11 gage, 1-½ inch long, 7/16 inch head nails	6" o.c.	6" o.c.
9	Square edge hardwood panel siding	3/8"	24" o.c.	6d common nails	4" o.c.	8" o.c.
	Shiplap edge hardwood panel siding	3/8"	16" o.c.	6d common nails	4" o.c.	4" o.c.
10	Reinforced cement mortar	1"	24" o.c.			

* BOCA National Building Code allows the use of only 2-M-1 and 2-M-2 particle board; SBCCI Standard Building Code allows the use of only 2-M-1 particle board.

Table No.9

BOCA National Building Code

Effective Peak Velocity Related Acceleration, A_v	Seismic Performance Category	Maximum Distance Between Interior Braced Walls (feet)	Maximum Stories (height) Permitted	Seismic Wall Bracing (4'-0" min. panel located at each wall corner and at 25 feet o.c. max or % of wall length as shown below).									
				Wall Panel Type in Top or Only Story									
				1	2	3	4	5	6	7	8	9	10
$A_v < 0.05$	A	No Limit	No Limit		NA	NA	NA	NA	NA				
$0.05 <= A_v < 0.10$	B	35	3 (40 feet)		X		X	20%	20%				
$0.10 <= A_v < 0.15$	C	25	2 (30 feet)		X		X	28%	28%				
$0.15 <= A_v < 0.20$	C	25	2 (30 feet)		20%		20%	36%	36%				
$0.20 <= A_v < 0.30$	D	25	2 (30 feet)		32%		32%	56%	56%				
$0.30 <= A_v$	D	25	2 (30 feet)		40%		40%	72%	72%				

BOCA National Building Code

Effective Peak Velocity Related Acceleration, A_v	Seismic Performance Category	Maximum Distance Between Interior Braced Walls (feet)	Maximum Stories (height) Permitted	Seismic Wall Bracing (4'-0" min. panel located at each wall corner and at 25 feet o.c. max or % of wall length as shown below).									
				Wall Panel Type in First of Two Stories or Second of Three Stories									
				1	2	3	4	5	6	7	8	9	10
$A_v < 0.05$	A	No Limit	No Limit		NA	NA	NA	NA	NA				
$0.05 <= A_v < 0.10$	B	35	3 (40 feet)		20%	72%	20%	36%	36%				
$0.10 <= A_v < 0.15$	C	25	2 (30 feet)		28%		28%	52%	52%				
$0.15 <= A_v < 0.20$	C	25	2 (30 feet)		40%		40%	68%	68%				
$0.20 <= A_v < 0.30$	D	25	2 (30 feet)		56%		56%	100%	100%				
$0.30 <= A_v$	D	25	2 (30 feet)		72%		72%	128%	128%				

BOCA National Building Code

Effective Peak Velocity Related Acceleration, A_v	Seismic Performance Category	Maximum Distance Between Interior Braced Walls (feet)	Maximum Stories (height) Permitted	Seismic Wall Bracing (4'-0" min. panel located at each wall corner and at 25 feet o.c. max or % of wall length as shown below).									
				Wall Panel Type in First of Three Stories									
				1	2	3	4	5	6	7	8	9	10
$A_v < 0.05$	A	No Limit	No Limit		NA		NA	NA	NA				
$0.05 <= A_v < 0.10$	B	35	3 (40 feet)		28%		28%	52%	52%				
$0.10 <= A_v < 0.15$	C	25	2 (30 feet)										
$0.15 <= A_v < 0.20$	C	25	2 (30 feet)										
$0.20 <= A_v < 0.30$	D	25	2 (30 feet)										
$0.30 <= A_v$	D	25	2 (30 feet)										

Table No.10

SBCCI Standard Building Code

Effective Peak Velocity Related Acceleration, A_v	Seismic Performance Category	Maximum Distance Between Interior Braced Walls (feet)	Maximum Height Permitted	Seismic Wall Bracing (4'-0" min. panel located at each wall corner and at 25 feet o.c. max or % of wall length as shown below).									
				Wall Panel Type in Top or Only Story									
				1	2	3	4	5	6	7	8	9	10
$A_v < 0.05$	A	N/A	No Limit	X	X	X	X	X	X	X			X
$0.05 \leq A_v < 0.10$	B	N/A	No Limit		X	X	X	X	X	X			X
$0.10 \leq A_v < 0.15$	C	N/A	35 feet		X	X	X	X	X	X			X
$0.15 \leq A_v < 0.20$	C	N/A	35 feet		X	X	X	X	X	X			X
$0.20 \leq A_v < 0.30$	D	N/A	35 feet		X	X	X	X	X	X			X
$0.30 \leq A_v$	D	N/A	35 feet		X	X	X	X	X	X			X

SBCCI Standard Building Code

Effective Peak Velocity Related Acceleration, A_v	Seismic Performance Category	Maximum Distance Between Interior Braced Walls (feet)	Maximum Height Permitted	Seismic Wall Bracing (4'-0" min. panel located at each wall corner and at 25 feet o.c. max or % of wall length as shown below).									
				Wall Panel Type in Second of Three Stories									
				1	2	3	4	5	6	7	8	9	10
$A_v < 0.05$	A	N/A	No Limit	X	X	X	X	X	X	X			X
$0.05 \leq A_v < 0.10$	B	N/A	No Limit		X	X	X	X	X	X			X
$0.10 \leq A_v < 0.15$	C	N/A	35 feet		X	X	X	X	X	X			X
$0.15 \leq A_v < 0.20$	C	N/A	35 feet		X	X	X	X	X	X			X
$0.20 \leq A_v < 0.30$	D	N/A	35 feet		40%	40%	40%	40%	40%	40%			40%
$0.30 \leq A_v$	D	N/A	35 feet		40%	40%	40%	40%	40%	40%			40%

SBCCI Standard Building Code

Effective Peak Velocity Related Acceleration, A_v	Seismic Performance Category	Maximum Distance Between Interior Braced Walls (feet)	Maximum Height Permitted	Seismic Wall Bracing (4'-0" min. panel located at each wall corner and at 25 feet o.c. max or % of wall length as shown below).									
				Wall Panel Type in First of Two Stories or First of Three Stories									
				1	2	3	4	5	6	7	8	9	10
$A_v < 0.05$	A	N/A	No Limit	X	X	X	X	X	X	X			X
$0.05 \leq A_v < 0.10$	B	N/A	No Limit		25%	25%	25%	25%	25%	25%			25%
$0.10 \leq A_v < 0.15$	C	N/A	35 feet		25%	25%	25%	25%	25%	25%			25%
$0.15 \leq A_v < 0.20$	C	N/A	35 feet		25%	25%	25%	25%	25%	25%			25%
$0.20 \leq A_v < 0.30$	D	N/A	35 feet		40%	40%	40%	40%	40%	40%			40%
$0.30 \leq A_v$	D	N/A	35 feet		40%	40%	40%	40%	40%	40%			40%

ICBO Uniform Building Code

Seismic Zone	Seismic Performance Category	Maximum Distance Between Interior Braced Walls (feet)	Maximum Height Permitted	Seismic Wall Bracing (4'-0" min. panel located at each wall corner and at 25 feet o.c. max or % of wall length as shown below).									
				Wall Panel Type in Top or Only Story									
				1	2	3	4	5	6	7	8	9	10
0	N/A	25 or 34 feet *	N/A	X	X	X	X	X		X	X	X	
1	N/A	25 or 34 feet *	N/A	X	X	X	X	X		X	X	X	
2A	N/A	25 or 34 feet *	N/A	X	X	X	X	X		X	X	X	
2B	N/A	25 or 34 feet *	N/A	X	X	X	X	X		X	X	X	
3	N/A	25 feet **	65 feet		X	X	X	X		X	X	X	
4	N/A	25 feet **	65 feet		X	X	X	X		X	X	X	

ICBO Uniform Building Code

Seismic Zone	Seismic Performance Category	Maximum Distance Between Interior Braced Walls (feet)	Maximum Height Permitted	Seismic Wall Bracing (4'-0" min. panel located at each wall corner and at 25 feet o.c. max or % of wall length as shown below).									
				Wall Panel Type in First of Two Stories or Second of Three Stories									
				1	2	3	4	5	6	7	8	9	10
0	N/A	25 or 34 feet *	N/A	X	X	X	X	X		X	X	X	
1	N/A	25 or 34 feet *	N/A	X	X	X	X	X		X	X	X	
2A	N/A	25 or 34 feet *	N/A	X	X	X	X	X		X	X	X	
2B	N/A	25 or 34 feet *	N/A	X	X	X	X	X		X	X	X	
3	N/A	25 feet **	65 feet		25%	25%	25%	25%		25%	25%	25%	
4	N/A	25 feet **	65 feet		25%	25%	25%	25%		25%	25%	25%	

ICBO Uniform Building Code

Seismic Zone	Seismic Performance Category	Maximum Distance Between Interior Braced Walls (feet)	Maximum Height Permitted	Seismic Wall Bracing (4'-0" min. panel located at each wall corner and at 25 feet o.c. max or % of wall length as shown below).									
				Wall Panel Type in First of Three Stories									
				1	2	3	4	5	6	7	8	9	10
0	N/A	25 or 34 feet *	N/A	X	X	X	X	X		X	X	X	
1	N/A	25 or 34 feet *	N/A	X	X	X	X	X		X	X	X	
2A	N/A	25 or 34 feet *	N/A	X	X	X	X	X		X	X	X	
2B	N/A	25 or 34 feet *	N/A	X	X	X	X	X		X	X	X	
3	N/A	25 feet **	65 feet		40%	40%	40%	40%		40%	40%	40%	
4	N/A	25 feet **	65 feet		40%	40%	40%	40%		40%	40%	40%	

* Spacing shall not exceed 34 feet on center in both the longitudinal and transverse directions in each story where the basic wind speed is less than or equal to 80 MPH, and spacing shall not exceed 25 feet on center in both the longitudinal and transverse directions in each story where the basic wind speed exceeds 80 MPH.
Exception: In one and two story Group R, Division 3 buildings, interior braced wall line spacing may be increased to not more than 34 feet on center in order to accommodate one single room per dwelling unit not exceeding 900 square feet. The Building Official may require additional walls to contain braced panels when this exception is used.

** Exception: In one and two story Group R, Division 3 buildings, interior braced wall line spacing may be increased to not more than 34 feet on center in order to accommodate one single room per swelling unit not exceeding 900 square feet. The Building Official may require additional walls to contain braced panels when this exception is used.

Table No. 12

CABO One and Two Family Dwelling Code

Seismic Zone	Seismic Performance Category	Maximum Distance Between Interior Braced Walls (feet)	Maximum Height Permitted	Seismic Wall Bracing (4'-0" min. panel located at each wall corner and at 25 feet o.c. max or % of wall length as shown below).									
				Wall Panel Type in Top or Only Story									
				1	2	3	4	5	6	7	8	9	10
0	N/A	N/A	N/A	X		X			X				
1	N/A	N/A	N/A	X		X			X				
2A	N/A	N/A	N/A	X		X			X				
2B	N/A	N/A	N/A	X		X			X				
3	N/A	N/A	N/A	X		X			X				
4	N/A	N/A	N/A	X		X			X				

CABO One and Two Family Dwelling Code

Seismic Zone	Seismic Performance Category	Maximum Distance Between Interior Braced Walls (feet)	Maximum Height Permitted	Seismic Wall Bracing (4'-0" min. panel located at each wall corner and at 25 feet o.c. max or % of wall length as shown below).									
				Wall Panel Type in First of Two Stories or Second of Three Stories									
				1	2	3	4	5	6	7	8	9	10
0	N/A	N/A	N/A	X		X			X				
1	N/A	N/A	N/A	X		X			X				
2A	N/A	N/A	N/A	X		X			X				
2B	N/A	N/A	N/A	X		X			X				
3	N/A	N/A	N/A			25%			25%				
4	N/A	N/A	N/A			25%			25%				

CABO One and Two Family Dwelling Code

Seismic Zone	Seismic Performance Category	Maximum Distance Between Interior Braced Walls (feet)	Maximum Height Permitted	Seismic Wall Bracing (4'-0" min. panel located at each wall corner and at 25 feet o.c. max or % of wall length as shown below).									
				Wall Panel Type in First of Three Stories									
				1	2	3	4	5	6	7	8	9	10
0	N/A	N/A	N/A	X	X	X	X	X	X	X	X	X	
1	N/A	N/A	N/A	X	X	X	X	X	X	X	X	X	
2A	N/A	N/A	N/A	X	X	X	X	X	X	X	X	X	
2B	N/A	N/A	N/A	X	X	X	X	X	X	X	X	X	
3	N/A	N/A	N/A		40%	40%	40%	40%	40%	40%	40%	40%	
4	N/A	N/A	N/A		40%	40%	40%	40%	40%	40%	40%	40%	

15. HOME BUILDERS CHECK LIST

A check list is provided for the home builder to use to determine whether important recommendations for seismic resistant design have been considered when constructing a residence. The builder should refer to the Guide for information on each of the items listed. This list should not be considered as being all inclusive.

FOUNDATIONS

1. Suitable and uniform ground conditions
2. Consistent foundation systems
3. Depth of footing below grade
4. Placing of reinforcing — lapped splices and at corners and intersections
5. Size, location, spacing and embedment of anchor bolts — type of washers
6. Location of hold-down anchor bolts or strap hold-downs
7. Treated lumber sill plates — 2x or 3x lumber

FLOORS

1. Square or rectangular plan
2. Diaphragm ratio limits
3. Perimeter blocking or rim joists nailed to sill plates or wall top plates
4. Splicing of collectors
5. Orientation of sheathing
6. Sheathing joints centered and nailed to common framing member.
7. Sheathing at openings blocked and nailed
8. Ties at split-level construction

SHEAR WALLS

1. Shear wall or panel in each exterior elevation
2. Four foot width of shear wall at each corner
3. Symmetrical pattern and balanced widths
4. Twenty-five feet maximum between shear walls/panel
5. Conformance with aspect ratios
6. Selection of wall bracing material
7. Wall sheathing materials on full height of wall
8. Consistent selection of wall sheathing material
9. Nailing of sheathing
10. Internailing of wall top plates
11. Hold-down at each end of shear wall/panel
12. Post or double studs at hold-downs
13. Attachment of hold-downs
14. Edge of sheet nailing to hold-down posts/studs
15. Blocking of all edges of sheathing not over studs
16. Bracing or shear panels at garage door openings

ROOFS

1. Square or rectangular plan
2. Diaphragm ratio limits
3. Perimeter supported on shear wall or panels

4. Edge joist, rim joist or blocking at perimeter nailed or fastened with clips to wall top plates

5. Blocking over interior shear walls

6. Orientation of sheathing sheets

7. Sheathing joints centered and nailed to common joists, rafters or blocking

8. Nailing of roof sheathing

9. Blocking and edge of sheet nailing at breaks in roof planes

10. No openings at corners

11. Blocking and edge of sheet nailing of sheathing around openings

12. Collector ties to shear walls or panels

MASONRY CHIMNEYS

1. Footing dowels each vertical reinforcing bar

2. Full height reinforcing bar — including extension above roof

3. Vertical reinforcing bar (minimum) each corner

4. Horizontal ties around vertical bars

5. Reinforcing bars fully embedded in grout

6. Blocking in roof framing to receive anchor straps

7. Strap anchor at each side of chimney

8. Strap anchor at each horizontal diaphragm bolted or nailed to framing

9. Nailing of Wood Structural Panel sheathing to block supporting strap

10. Blocking and nailing around opening in roof for chimney

11. No heavy veneer at adjacent shear walls or panels

CONCRETE MASONRY

1. Square or rectangular floor plan

2. Continuous wall footing — reinforcing and wall dowels

3. Corner shear walls of proper dimension

4. Bond pattern

5. Horizontal/vertical reinforcing

6. Horizontal chord bars — spliced for continuity

7. Vertical reinforcing bars at corners

8. Horizontal bars lapped and tied at corners

9. Reinforced cells grouted

10. Wall anchor straps attached to floors/roofs

11. Sheathing nailed to tops of ledgers at face of wall

CLAY MASONRY

1. See CONCRETE MASONRY

2. Reinforcing in cavity or cells fully embedded in grout

3. Metal wall ties or joint reinforcing

4. Spacing of wall ties — horizontal/vertical

5. Empirical rules for wall thickness
6. Height to thickness ratio – shear walls
7. Running bond – unreinforced

MASONRY AND STONE VENEER

1. Firm backing – Wood structural Panel
2. Tooled mortar joints
3. Concrete or steel foundation/support
4. Bond pattern
5. Lath attached with non-corrosive nails/screws
6. Space provided between veneer and backing (may be filled)
7. Wire tie anchors – galvanized
8. Joint reinforcing

16. APPENDIX

TYPICAL REGULAR FLOOR PLANS
FOR
EARTHQUAKE RESISTANCE

FLOOR PLAN - ONE STORY
(NEARLY SQUARE FLOOR PLAN)

FLOOR PLAN - ONE STORY
(NEARLY SQUARE FLOOR PLAN WITH ATTACHED GARAGE)

FLOOR PLAN - ONE STORY SPLIT LEVEL
(REGULAR/OR RECTANGULAR FLOOR PLAN)

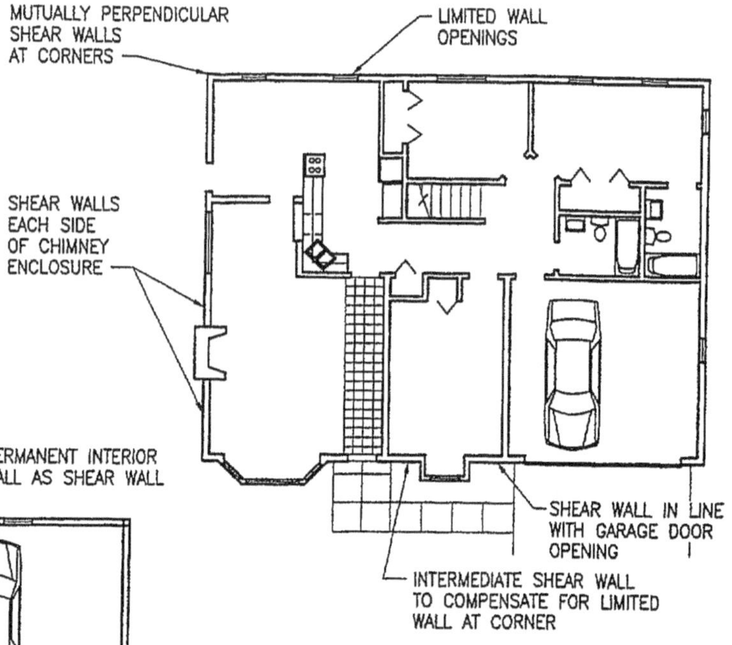

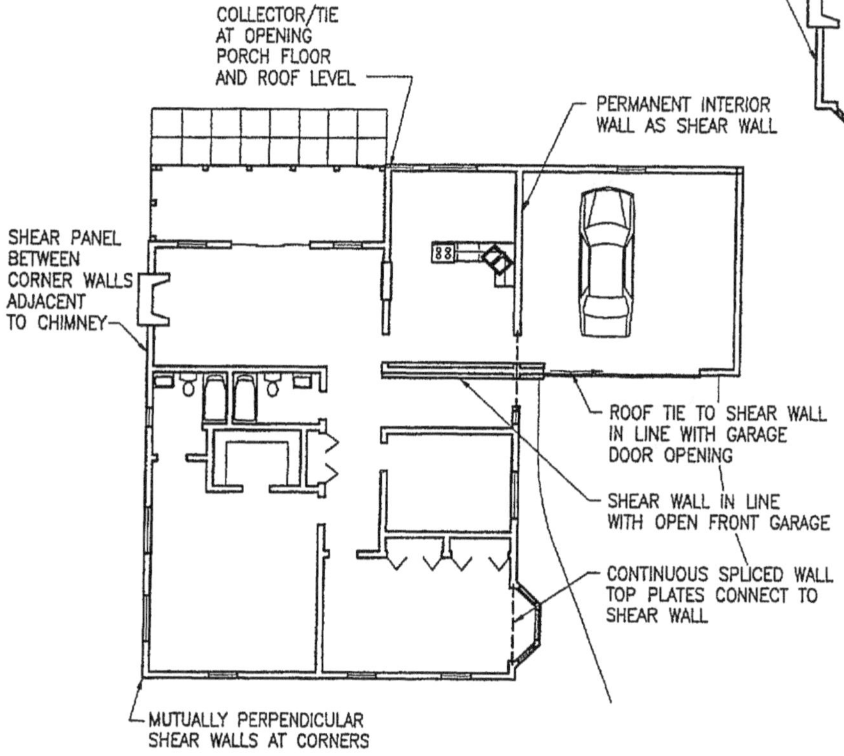

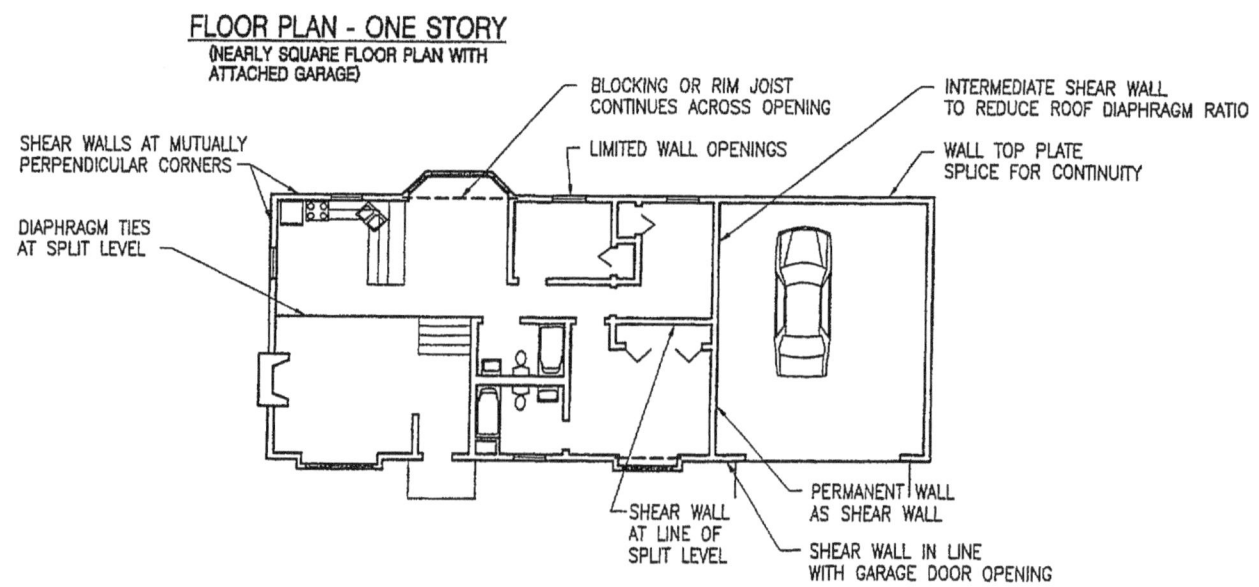

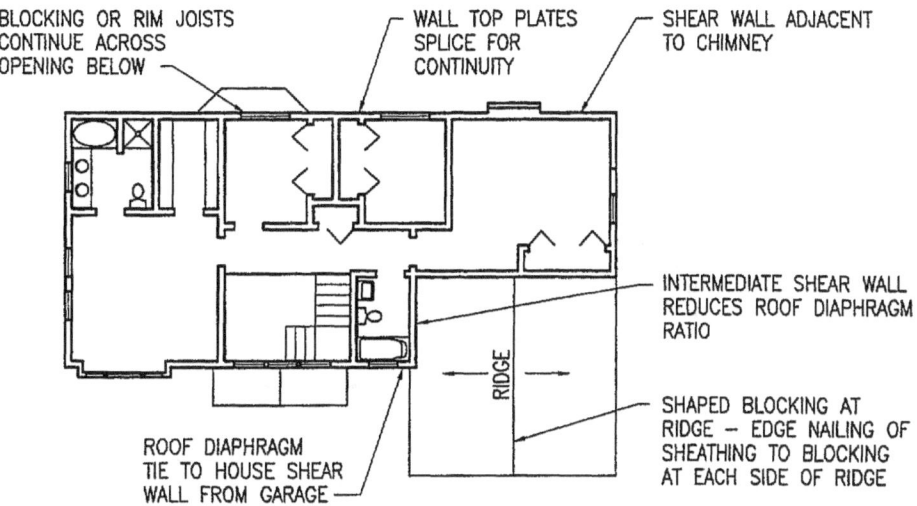

PLAN – SECOND STORY

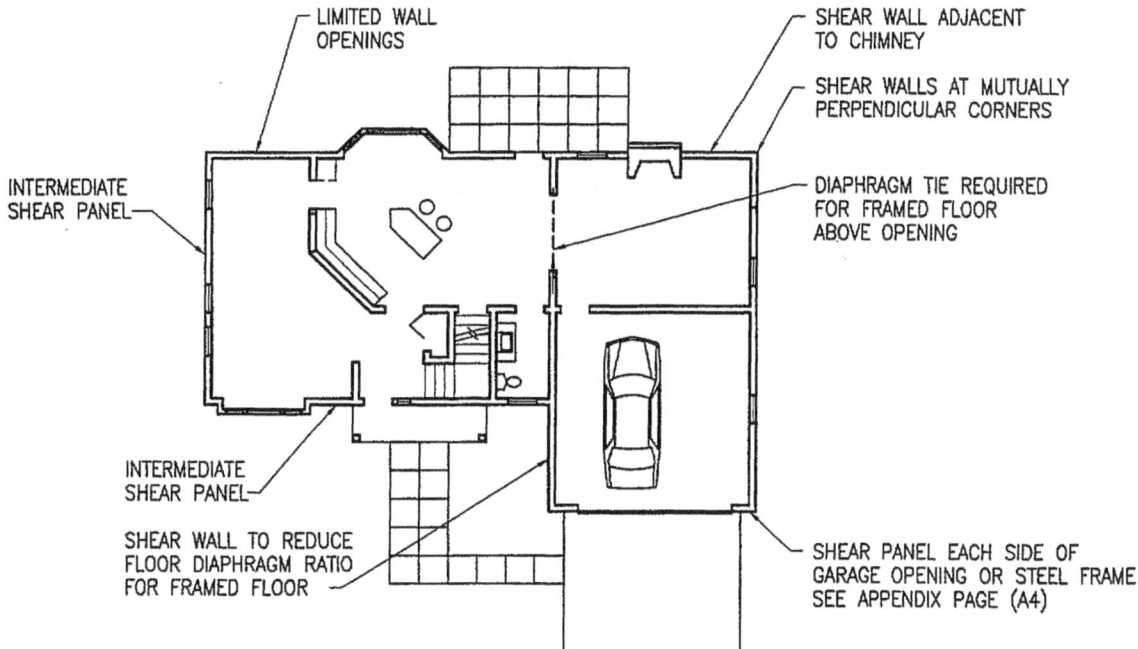

PLAN – FIRST STORY

FLOOR PLANS - TWO STORY
(NEARLY RECTANGULAR FLOOR PLAN)

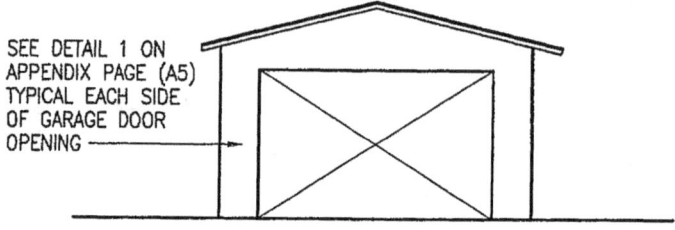

OPEN FRONT GARAGE

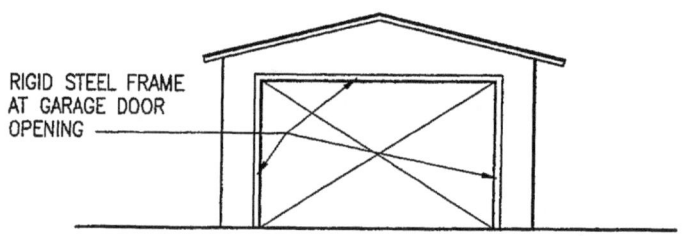

OPEN FRONT GARAGE
(OPTIONAL)

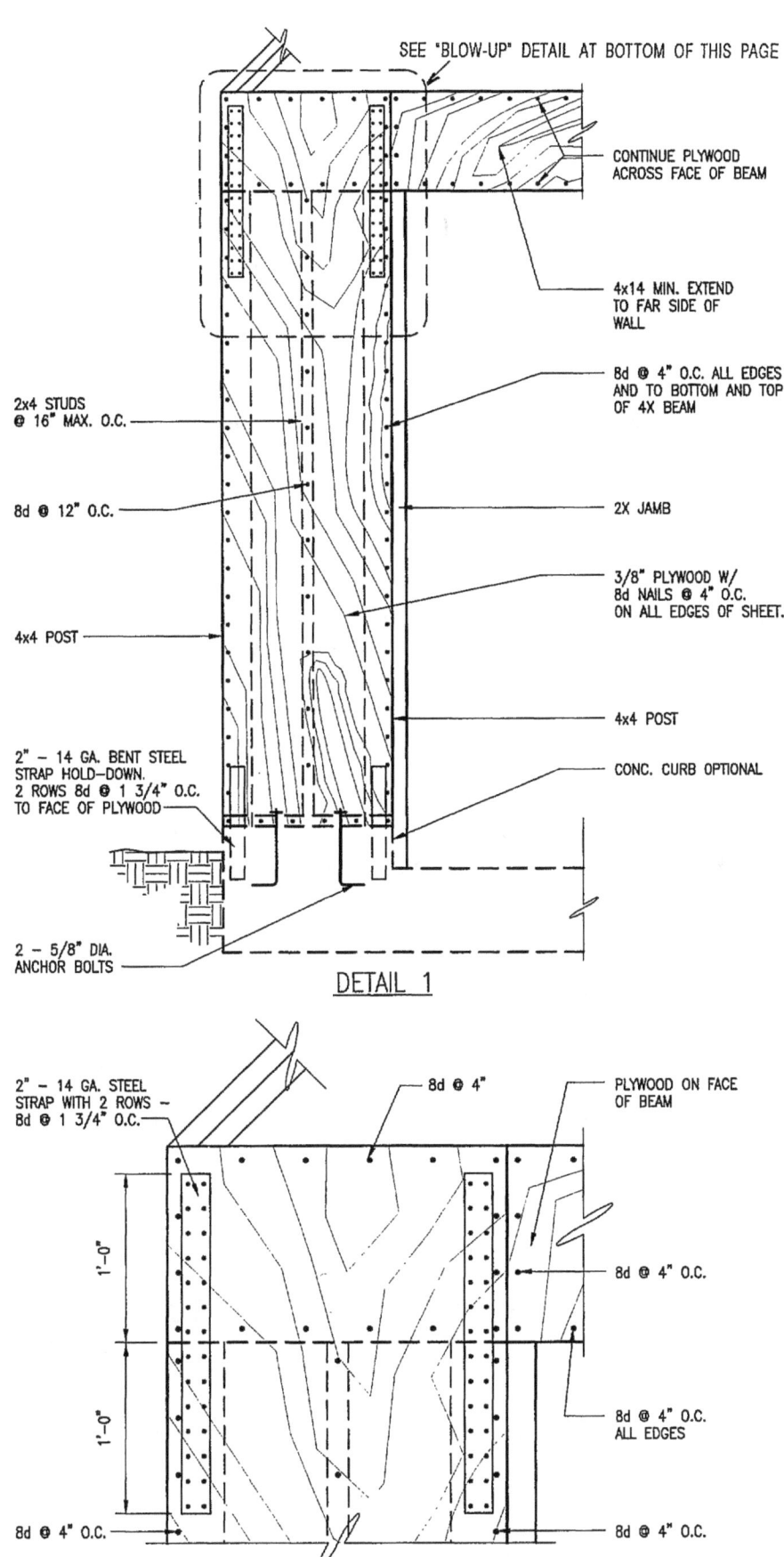

SPECIAL GARAGE FRONT WALL DETAIL FOR SINGLE STORY GARAGES

(A5)

ACKNOWLEDGMENTS

SOHA Engineers gratefully acknowledges the help and guidance of the following members of the team of professionals who assisted in and contributed to the preparation of this Guide. William W. Stewart, FAIA (Stewart-Schaberg/Architects), Maury Power (Geomatrix) and Shyam Choudry (Housing and Building Technology).

We also appreciate the care and direction provided by the following individuals associated with the Federal Emergency Management Agency — Kenneth Sullivan, Michael Mahoney and John Gambel.

In addition, SOHA Engineers and the Federal Emergency Management Agency wish to extend their sincere appreciation to the following persons and their affiliation for providing a review of the Guide prior to final publication. Their generous and insightful diligence has contributed importantly to the usefulness and technical accuracy of the material presented throughout the text.

Susan Dowty	International Conference of Building Officials, ICBO, Whittier, CA
Philip Line, PE	American Forest & Paper Products
Dr. Dan Dolan, PE	Virginia Polytechnic Institute and State University
Kelly E. Cobeen	GFDS Structural Engineers, San Francisco, CA
Mark Hogan	National Concrete Masonry Association
Hank Martin	American Iron & Steel Institute (AISI)
Dr. Phillip Lowe	Intech Solutions, Inc., Gaithersburg, MD
J. Gregg Borchelt, PE	Brick Institute of America
Ken Bland	American Forest and Paper Association
Ken Ford	National Association of Home Builders
Robert McClure	Building Officials and Code Administrators (BOCA)
Richard A. Vognild, PE	Southern Building Code Congress International, Inc.
Kenneth Anderson	American Plywood Association (APA)

www.ingramcontent.com/pod-product-compliance
Lightning Source LLC
Chambersburg PA
CBHW081219230426
43666CB00015B/2797